AF777925

A Physical and Spiritual Reality

An Essay Connecting Theology and the Life Sciences

Daniel J. Price

SUMMERLAND PUBLISHING

Salt Lake City, UT

SUMMERLAND PUBLISHING

ISBN: 978-1-7375730-1-2

Library of Congress #2021917364

© Copyright 2021 by Daniel Price

All rights reserved. No part of this book may be reproduced or transmitted in any form or by any means, electronic or mechanical, including photocopying, recording, or by any information storage and retrieval system, except in the case of brief quotations embodied in critical articles and reviews, without prior written permission of the publisher: Summerland Publishing, 3181 E. Elgin Drives, Salt Lake City, UT 84109.

Layout and Design by Pizzirani Consulting

Cover details: Three-dimensional structure of cytochrome c (green) with a heme molecule coordinating a central Iron atom (orange). PDB id, 1HRC, Bushnell et al., "High-resolution three-dimensional structure of horse heart cytochrome c." J Mol Biol. 1990 Jul 20;214(2):585-95. PubMed PMID: 2166170. (see Vossman, Wikimedia Commons, last updated [09-13-2011], [https://commons.wikimedia.org/wiki/File:Cytochrome_C.png]).

Dedication

This essay is dedicated to my high school biology teacher, Alex Easton. Alex, through his enthusiasm for the life sciences, was the first to encourage me to enter college as a biology major. In his teaching and mentoring of students, Alex conveyed that electricity which is the essence of discovery in the natural sciences.

Table of Contents

Introduction

Over the last two centuries, we have seen tremendous advances in the scientific understanding of our world, particularly in the area of the biological sciences. Advances in biochemistry and molecular biology have probed in great detail the molecular nature of how we function physiologically and how we interact with our environment. This expanded understanding of our world has led to advances in medicine that have greatly improved and extended our lives. Our understanding of the genetic nature of life has added greatly to our understanding of how present life forms arose from previous primordial creatures. However, many questions have gone unanswered about how God might have initiated such a creation and how he* interacts with his creation in our present day. This essay came to be out of my experiences growing up and throughout my adult life as a biological scientist and as a spiritual person. The major theme is to harmonize the study of the natural sciences with what we have come to know about the monotheistic God who has been revealed through Judeo-Christian history. Furthermore, it is about connecting those spiritual and physical realities, including events involving our exploration of the natural sciences and God's intervention into our world. The intent is to spur a conversation about how we regard the spiritual and the physical phenomena of our lives. Are they both evidential realities? Are they mutually exclusive, or do they overlap? Do we need both of these to fully appreciate the world that we live in? These are some of the questions that I intend to address in this essay. I encourage you, the reader, to engage with them and debate them, and I hope that in so doing you will come to a greater understanding of how they apply to our lives.

This essay heavily reflects a worldview involving a transcendent God who superintends a physical world governed by understandable and testable physical laws. Such a worldview has tremendous implications for how we live in this world, how we treat others, and how we view the universe as a whole. While many from the scientific community—notably Richard

Dawkins, Stephen Hawking, and Lawrence Krauss—and others have derided such a worldview as impossible, they have never been able to effectively prove that this God is absent or that he does not participate in our physical world.

It is not the purpose of this essay to defend theism over atheism in terms of their relationships to the natural sciences. This argument has been made by a number of competent authors including Francis Collins, Karl Giberson, Dorothy Sayers, C. S. Lewis, and others. It is important not to minimize the importance of these discussions about theism and atheism in regard to the natural sciences. Many respected scientists have concluded that to include God in the equation of the scientific endeavor is to either weaken the conclusions that science makes about the universe or to totally prevent any objective understanding of natural laws. In his book *The Language of God*, Francis Collins answers this criticism:

> *In this modern era of cosmology, evolution, and the human genome, is there still the possibility of a richly satisfying harmony between the scientific and spiritual worldviews? I answer with a resounding yes! In my view, there is no*

> *conflict in being a rigorous scientist and a person who believes in a God who takes a personal interest in each one of us.*[1]

Many would like a world in which God did not take such a personal interest, because that would make the doing of science a little bit easier. We would not be so worried that God would interfere with our exploration of the natural world or that he would randomly throw in a wrench that would prevent us from coming to the right conclusion. However, because God is truly transcendent, we can postulate that he resides in a realm that is sufficiently outside the physical, measurable universe and that he is not intent on interfering with our humble efforts to understand how things work in our physical universe. So, he can intervene in our daily lives in innumerable ways, but we never have to worry that the apple falling from the tree will not reproducibly follow Newton's laws of motion. C. S. Lewis, in his little book *Miracles*, asks: "Would not an emergent God give us all we need? Is it really necessary to bring in a super-natural God, distinct from and outside the whole interlocked system?" He goes on to reason:

> *But I am afraid that will not do. It is, of course, possible to suppose that when all the atoms of the universe got into a certain relation (which they were bound to get into sooner or later) they would give rise to a universal consciousness. And it might have thoughts. And it might cause those thoughts to pass through our minds. But unfortunately, its own thoughts, on this supposition, would be the product of non-rational causes and therefore, by the rule which we use daily, they would have no validity. This cosmic mind would be, just as much as our own minds, the product of mindless Nature. We have not escaped from the difficulty; we have only put it a further stage back. The cosmic mind will help us only if we put it at the beginning, if we suppose it to be, not the product of the total system, but the basic, original, self-existent Fact which exists in its own right. But to admit that sort of cosmic mind is to admit a God outside Nature, a transcendent and supernatural God.*[2]

So, the argument is that we cannot go halfway. We cannot argue for a God who is the watchmaker and sets the world spinning to go according to predetermined laws and remains separate from it. We cannot imagine a God who acted in creation but is now totally uninterested in its fate. Thus, we must conclude that God is a God of natural laws but is also capable of interacting with his world in ways that include personal intervention into human affairs. In this essay, I am taking more or less as a starting point the existence of a transcendent God who has established a world that is understandable in terms of testable, defined physical laws. Despite the lack of detail here, the reader is encouraged to dig deeper into these other foundational works on faith and the natural sciences.

If other authors have so effectively defended the harmony of theism and the study of the natural sciences, what need is there for further enquiry? The point is that, as will be shown in the later chapters, looking at applications in the life sciences, we can see many reasons to include God in the equation. In the following chapters, I will highlight a number of cases where consideration of God and his involvement in the life sciences not only is consistent with an understanding of the physical world but also leads to a greater understanding of the physical world.

---

* While masculine pronouns are used throughout the text in reference to God, it is understood that many scholars are more comfortable with referring to God as neither male nor female.

Chapter 1:

A Student of the Natural Sciences Finds Jesus

Raise your eyes on high
and see who has created these stars,
The One who brings out their multitude by number,
He calls them all by name;
Because of the greatness of His might and the strength of His power,
Not one of them is missing.

Isaiah 40:26[3]

As a youth growing up in Long Island, New York, I would probably have been considered introspective or even antisocial. For recreation, our family (my parents, two brothers, and one sister) would spend a lot of time sailing our boat, *The South Wind*, on the Great South Bay. Some of those outings were at night, as we returned from the Fourth of July fireworks, or some extended overnight as we sailed on the bay. I distinctly remember sitting out near the bow of the boat and gazing up at the stars. In those days, the Milky Way was very much visible, and I was taken aback by the vastness of the universe as well as the quietness of the dark water lapping against the side of the boat. Later in my teens, I took up black-and-white photography as a hobby and extended my appreciation of the natural world. Figure 1.1 shows a photo that I shot of the sun breaking through a dense forest. We used this photo as a Christmas card in the early 1970s with the quote of John 1:4: “In Him was life, and the life was the Light of mankind.” So, even in my younger years, I had an appreciation that God was at work in nature. Our parents brought us up in the Congregational Church (United Church of Christ, or UCC), so seeing God at work in nature was for me something that made total sense. I just didn’t know the details.

Figure 1.1. Sun breaking through a dense forest (photo by the author).

I grew up in a family that was filled with science and music. My father was an accomplished nuclear physicist who spent most of his career investigating nuclear reactor safety and promoting nuclear energy. As a young army private in 1945, he was present at the first nuclear bomb test in Alamogordo, New Mexico. From a letter that he wrote to his father after this test, it was clear that this experience was formative for his life and influenced him to pursue a career in nuclear physics. My mother was an organist at our church and also started a local fine arts organization that we were all involved with as a family.

In my early years, not a lot of my father's interest in science rubbed off on me. My math skills were borderline. I fell asleep in chemistry class. In biology, I considered the dissection of the fetal pig, pickled in formaldehyde, to be one of the most distasteful experiences one could ever imagine. However, I did manage to buckle down in my last two years of high school and significantly improved my GPA. As a senior at Bayport High School, I signed up for AP Biology. Our teacher, Alex Easton, was one who really got involved in his subject and had an innate curiosity that he managed to convey to a class of not-always-attentive teens. I remember distinctly watching for the first time the live dissection of a frog, seeing the beating heart and the dark, glistening liver. This was not the formaldehyde-pickled dissection that I had seen before. I discovered that there was a lot going on in the field of biology—microscopically, molecularly, and physiologically—that I had not noticed before.

Young people enter college for a variety of reasons—to further a career, to fill the time between the teens and the twenties, to get away from parents, or possibly because there is nothing better to do. I entered Johns Hopkins University in fall 1972 because I wanted to understand how living things tick. That sounds a little altruistic, but I think that it was actually the case. I obviously had thoughts of a future career, but that was not first in my mind. I was still driven by that eager curiosity for the natural world that Alex Easton had planted in my mind in AP Bio. I entered as a biophysics major, possibly as a subconscious nod to my dad that I would take some physics courses. I did take those courses, but later ended up switching to biology as a major for lack of mathematics expertise. I do not regret starting out in biophysics, because it exposed me to some really significant areas relating to the three-dimensional structure of protein and nucleic acid. We learned these lessons from professors who had been at Johns Hopkins when the field was born. The 3-D structure of DNA was solved by Watson and Crick in 1953, and the first globular protein structures were solved in the late 1950s. I was impressed to see these structures come to life in three-dimensional space with the aid of a stereo viewer. It was like virtual reality at the molecular level. Not only did these proteins and nucleic acids have defined 3-D structures, but they also functioned like little molecular machines operating according to physical and chemical principles. We also learned about the basics of molecular biology—how the nucleic acid code perpetuated itself and was translated in cellular proteins. Many first-year students find courses to be either a repeat of what they had in high school or just a list of prerequisites for courses that they were really interested in. I felt as if a new world had been opened up, showing how things operate on a molecular level.

One of the things that we did between terms was to watch a showing of Stanley Kubrick's *2001: A Space Odyssey* on campus. We stayed up all night discussing the implications that this science fiction flick had for real science: Could there be a real Hal the computer? What was the significance of all those monolithic signs of the next frontiers for the human race? Was there another dimension to the world that we lived in? My friends and I would spend hours thinking of the implications of this movie intertwined in all the science we were learning in our classes. We got the key to the astronomy observatory and checked out the planets to see if they were in the correct orbits (they usually were). We also occasionally pointed the telescope to the girls' dormitory to check out other heavenly bodies. While I made little social progress in my first year, I did open up a new world of reality at the molecular level.

Summer after my first year at Hopkins was notable for two things. First, I did a summer job at my dad's place of employment, Brookhaven National Laboratory. I spent time in the laboratory of Dr. Bill Siegelman in the Biology Department studying plant chromophores. In addition to seeing science close-up, we would also break religiously for coffee twice a day. There was a ritual making of the coffee in the lab that involved using all labware: a Bunsen burner, Buchner funnel, and vacuum flask. We also ate lunch in the lab and had a big jar in which people would bring in their garden vegetables to make Bill's special version of kosher pickles. In 1973, there were minimal lab safety concerns.

The other notable event that happened that summer was that my sister, Bev, invited my brother

Dave and me to a tent meeting hosted by a local church. I don't remember the context or why this was going on at that time, but it ended up being fairly significant in my life. After a short message from the Bible, the speaker mentioned that I could have a personal relationship with Jesus if I repented of my sins and accepted him. I had never heard of anything like this before. I wanted to find out more, so I went forward and accepted Jesus into my heart. My brother Dave did the same. Later I thought, "What have I done? I've accepted Jesus, who claims to have been raised from the dead, but I also believe in science, that things work in an ordered way. How does that work?"

I spent the next two years taking courses, eating bad dining hall food, and debating in my mind how to reconcile Christ's resurrection and what I was learning in the sciences. My initial conclusions were that nature itself was much more complex than we could understand, and that the presence of that complexity might explain such supposedly miraculous occurrences. I would always bounce my ideas off my maternal grandmother, who was a strong Christian, in letters that I wrote to her. She didn't have any scientific training or impressive educational degrees. However, she did have eighty-plus years of spiritual insight and experience in what made people tick. She would always encourage me, but also was honest to give her unvarnished opinion.

Toward the end of my junior year, I concluded that it was God who was much more complex than we could ever understand and that, in that respect, he was able to both establish physical laws and intervene in human affairs. The last physics course that I took at Hopkins ended with a discussion of fundamental particle theory. At the time, physicists used versions of string theory to explain how a fixed set of subatomic particles (quarks, leptons, photons, bosons, and gluons) and four fields (electromagnetic, weak, strong, and gravity) could explain the known universe (commonly known as the Standard Model). While I never pursued a deeper understanding of how string theory explained the goings-on at the subatomic level (probably to the disappointment of my father), it was clear that when you get to this level, the physical reality is far more complex than even what we observe at the molecular level. One of the implications of string theory is that there is the possibility of higher dimensions of space that we may not even have access to in our physics experiments. Toward the end of my physics course in spring 1975, I was walking to one of the physics lectures on string theory, and I put together some disconnected ideas—that a transcendent God could occupy some other dimension and actually intervene in the three dimensions of our physical universe. Furthermore, it seemed that God could involve himself in our physical universe and be responsible for things like miracles and other interventions in our lives. As I walked across the Gilman quadrangle toward my physics lecture, I had this picture in my mind of a giant theater in the round, and on the stage was the entire universe with all the spotlights shining on it. Outside of the stage was the realm that God inhabited, a sort of other-dimensional world. It was a grand sight, and I thought, maybe this is the larger world that the transcendent God dwells in, surrounding his creation. Would it be too presumptuous to suppose that God could interact with our physical universe in a personal way? I saw no reason why he couldn't. In fact, it made perfect sense that he would do so. I started to understand that a truly transcendent God could be responsible for both spiritual and physical realities. He exists outside of our limited three dimensions, the ones that I had learned so much about in my classes. However, he superintends all that goes on in those three dimensions. As the apostle Paul wrote to the Colossians, "He is before all things, and in Him all things hold together" (Colossians 1:17). Does that mean that God plays the part of a puppeteer, controlling the events on the stage of life? Absolutely not. However, as the God who creates and upholds this physical world, he maintains the physical laws, but also reserves the right to intervene in human affairs. He grants us free will and the opportunity to choose our own path, but he provides the path. In fact, he could even manage a miracle such as the resurrection of Christ, and at the same time maintain the physical laws.

So, the ability to reconcile the God of scientific law and the God of the Bible was in the end made possible by the greatness of God himself. What I arrived at was not a truce between two warring ideas, but rather a firm foundation on which I could do science and also trust that God was who he claimed to be.

There is a misconception concerning spiritual revelation that we get an epiphany from God, and everything going forward is all of a sudden made totally clear. This was not the case for me in my spiritual development and my development as a life scientist. I have spent the last more than forty years getting to know God through his word, and his creation through the scientific method. I have seen how the same God has given us both spiritual and physical laws that are revealing of his creation. These laws are not in conflict. They are merely in different realms.

Following my graduating from Johns Hopkins in 1976, I headed to a graduate program in biomedical science at the University of Tennessee (UT), Oak Ridge. Within two years, I transferred to a graduate level biochemistry program at UT, Knoxville, a short distance from Oak Ridge. Shortly after transferring to UT, Knoxville, I connected with some Christian friends that I had met in Oak Ridge, and started attending their church. We had a lot in common, spiritually and academically, but it became apparent that there were some differences as well. At one point they asked me whether I had ever heard of the human footprints found in dinosaur footprints. I said that I hadn't, so they introduced me to a book by John Whitcomb and Henry

Morris entitled *The Genesis Flood.*[4] This is a book that promotes a literal interpretation of the Bible, including the universal world-wide flood of Noah, and a literal six twenty-four-hour days of creation in Genesis 1. The church was equally strong in promoting these views. I should say at the outset that these types of interpretations are common among conservative evangelical Christian churches. I should also say that I respect them for being true to their convictions in keeping to a biblical standard.

I did a bit of research on this issue. I spoke with a professor in the zoology department at the University of Tennessee, and he made the point that he was a committed Christian, but he also accepted the theory of Darwinian evolution. Furthermore, he made the point that all the Christians in the zoology department also accepted the theory of evolution. I spent a lot of time debating in my own mind how the spiritual and physical laws are revealed in God's creation. While I respected those who believed that God's word was a detailed, literal description of the creation events, I could not accept this view. In the end, I had simply come to a different interpretation based on what God had revealed to me about his word and the physical creation. I had an equally strong commitment to biblical inspiration, but with the view that God has given us the scientific method as a way to know his physical laws and to appreciate his physical universe. Moreover, I would reason that by taking an extremely literal view of God's word, we are denied the opportunity to know God in his physical creation. This opportunity to know his world through the scientific method is not only beneficial to us; it is also essential to appreciating the majesty of his creation as we will see in the coming chapters.

Chapter 2:

The Source of the Structure—
Man Made in the Image of God or God Made in the Image of Man?

Before we address the nature of structures in the physical or spiritual realms, we must first enquire into where these structures came from. Often, an understanding of the nature of physical or spiritual interactions has been defended without an examination of the worldview that the understanding is based on.

The atheistic world view has largely originated from the presupposition that religion was derived from human imagination. It has been hypothesized that from the beginning the fully functional human mind had the capacity of imagination that would envision worlds and superhuman beings with no basis in reality. Such notions would be founded in dreams or visions that the atheistic worldview considered to be merely the afterthoughts or echoes of the human mind. In his book *Humans: From the Beginning; From the First Apes to the First Cities*, Christopher Seddon promotes the theory that religion developed from the capacity of the human mind to imagine superhuman beings and worlds beyond our physical world. He suggests that "anthropomorphism [is] the attribution of human characteristics to non-human things" and then poses the following question about religion as a mental construct in human development: "If we accept that religion is indeed a product of the way the human brain works, the obvious question is why? Is religion useful, or is [it] a 'misfiring by-product' of something more useful, as Richard Dawkins has suggested?"[5]

Seddon suggests, drawing from the work of Robin Dunbar,[6] that the capacities for storytelling and religion have evolutionary advantages that contribute to social bonding and cohesion among these socially advanced groups. The argument is that religion plays an advantageous function in groups of humans that leads to further social development. However, Seddon points out that such functionalism is not necessarily a requirement for social development, and the argument for this explanation of universal religious origins coming from human mental capacity may not be well founded. As support for the theory-of-the-mind origin of religion, Seddon points to the work of David Lewis-Williams in his book *The Mind in the Cave: Consciousness and the Origins of Art*. According to Lewis-Williams, shamanistic religion came about in Paleolithic times through altered states of consciousness induced by meditation, use of psychotropic substances, and other ritual practices.[7]

This worldview of religion coming from the complexity of the human mind is in stark contrast to the worldview that there are demonstrable realms and divine beings inaccessible to our physical senses. Physicists have hypothesized that there may be such realms in either multidimensional space or multiverses that are essentially beyond the reach of our senses.[8] These other realms have been invoked in support of fundamental particle physics and string theory. They have also been related to theories of the early universe and the so-called big bang. The point to be made is that while the theistic worldview is not synonymous with these views of physicists, it is certainly compatible with these views.

The theistic worldview can be summarized at least in part by certain statements coming from the Old Testament. In the opening of Genesis 1 we have the statement "In the beginning God created the heavens and the earth." This implies that God is a knowable being and independent of man's imagination. It also implies that God exists in a realm of a higher order than that of our physical universe. The proposition that God created the world rather than the world creating itself through physical interactions has been widely debated and will not be addressed here.

Later in Genesis 1, we have the statement "God created man in His own image, in the image of God He created him; male and female He created them" (1:27). Again, it is implied that God is a demonstrable, all-powerful being, existing in a higher realm, and that he was active in creating the first humans. Importantly, Genesis 1 does not say how God created these humans, but it does say that they were created in his likeness. This is in sharp contrast to the idea that man created the notion of an all-powerful God formulated in a human image.

To settle the apparent differences between the atheistic and theistic worldviews, it appears that there is little physical evidence to support either. However, we can ask, "Is there any historical evidence to support either view?" We see limited documented evidence of the mind formulating religious concepts through dreams or use of psychotropic drugs. In the New Testament scriptures, we find no evidence of shamanistic practices or of higher forms of consciousness leading to religious experiences. For instance, we have the account of Jesus' transfiguration, where Peter, James, and John experience his supernatural appearance with the prophets Moses an Elijah (Matthew 17:1–8). There is no indication that this vision was brought about by meditation or by any shamanistic practice. Later on, Peter testified the following about these events: "For we did not follow cleverly devised tales when we made known to you the power and coming of our Lord Jesus Christ, but we were eyewitnesses of His majesty. . . . and we ourselves heard this declaration made from heaven when we were with Him on the holy mountain" (2 Peter 1:16–18). So, it was clear to Peter, and probably the other eyewitnesses, that this was a real event, and not a dream or some higher level of consciousness.

This and many other supernatural events documented in the New Testament (for example, the resurrection of Jesus, the raising of Lazarus, the appearances of Jesus to his disciples) seem to confirm that those who experienced these things believed that they actually happened. As historical records, these accounts support more heavily a theistic view of a transcendent God acting from a higher-than-physical realm.

Chapter 3:

God in the Spiritual Realm Reflected in Our Physical World

The heavens tell of the glory of God;
And their expanse declares the work of His hands.
Day to day pours forth speech,
And night to night reveals knowledge.
There is no speech, nor are there words;
Their voice is not heard.

Psalm 19:1–3

In this verse from the Psalms, David indicates that not only is God the creator of the physical universe, but his glory is of a higher order than what we see in the physical heavens. Why did not David say that "nature tells of the glory of God"? The implication is that the realm of the spiritual heavens, where God dwells, is of a far higher order than even the physical heavenly realm with its enormous physical expanse (i.e., millions of light years across). But, if we could imagine this spiritual realm where God dwells, could we picture it in our minds? Many would say that this is an impossibility, but the psalmist says that we can see a reflection of this spiritual reality in the physical heavens, the world that we see when we look through the most powerful telescopes.

The year 1968 changed the way we viewed the heavens. As Andrew Chaikin has described in his book, *A Man on the Moon*,[9] 1968 had been a bad year for America, including assassinations of two national figures, the war in Vietnam, and riots in major cities. In December 1968, three American astronauts circled around the dark side of the moon on Christmas Eve and saw a sight that has become known as earthrise, the planet earth floating above the barren landscape of the moon in the blackness of space. They took photos that have become iconic of a new view of the cosmos, our home planet as viewed from its next closest relative (see figure 3.1). Although we had all seen images of the blue planet from earth orbit, we had never seen this view from the perspective of another celestial body. Later that day, the astronauts gave a live televised broadcast to half a billion people on earth that had tuned in to hear the experiences of three men circling the moon. They described vast craters below them and monolithic moonscapes as contrasted to the multicolored "grand oasis,"[10] the world that they had come from. As a final message, they read from the first chapter of Genesis: "In the beginning, God created the heaven and the earth." The verses were appropriate in that they captured the wonder that these astronauts experienced at viewing their home planet from such a perspective. In the monotony of doing their tasks to get to and from the moon, they had experienced a new point of view that no other humans had ever experienced. There was a new sense of brotherhood for all mankind because we were all seen as occupants of the same planet. It was a new sense of our place in the universe, a place not of preeminence, but of humility. We occupied such a small corner of the universe, but in such a corner, there was this "grand oasis."

Figure 3.1. Bill Anders/NASA, Earthrise (see Wikipedia, s.v. "Earthrise," last edited August 5, 2021, 21:14).

Of the American astronauts, twenty-four made the journey to lunar orbit between 1968 and 1972, and twelve of these astronauts walked on the surface of the moon. Each astronaut had his own story of how the journey to the moon affected them. Chaikin has told the stories of these lunar astronauts in *A Man on the Moon*. Two of the astronauts, Ed Mitchell (*Apollo 14*) and Jim

Irwin (*Apollo 15*), were noteworthy in that their journeys led them to spiritual experiences.

According to Chaikin, Mitchell experienced psychic phenomena that he attributed to information that permeated the universe. Mitchell felt that this information represented the "accumulated experiences of all matter, including living organisms throughout history." Mitchell further posited that "a person who is sensitive to this information—which exists, he says, as a form of energy—can experience anything from clairvoyance to telekinesis." Mitchell's concept of divinity was that it was the "intelligence existing in the universe" and that it was something inherent to the universe.[11]

Jim Irwin was a person of faith and a Christian before his *Apollo 15* journey, but his faith was deepened as a result of experiences on the journey. Irwin believed in the power of prayer. He also committed technical issues to God in prayer, and he saw answers to many of these prayers. It is indeed unusual for an astronaut trained in the technical issues of engineering and space flight to unite these nuts-and-bolts issues with a faith in God. However, Jim Irwin was such a person. As he walked on the surface of the moon, he was reminded of a verse from Psalm 121 (King James Version): "I will lift up mine eyes unto the hills, from whence cometh my help. My help cometh from the Lord." Irwin would later recount how while on the moon, he would find a white rock on Spur Crater, "presented on its own pedestal as if on an outstretched hand." Irwin's experiences on *Apollo 15* were a confirmation, according to Chaikin, that "God is alive and available everywhere in the universe."[12]

Irwin's and Mitchell's experiences were different, but they both witnessed the presence of God on their lunar journeys, and these experiences were apparently inherent in the world that they found in the depths of space. These experiences were not concepts that they had brought with them from Earth, or musings of their own minds. This is important in light of what was discussed in the last chapter, that either God is an invention of our own minds, or he exists independently of the human experience and is omnipresent in the universe around us. These experiences of Irwin and Mitchell support the idea that God exists in the far reaches of the universe, independent of our own human imaginations.

We now address the question of whether God exists beyond the domain of this universe, and if so, what is the relation of this "other universe" to our own universe. In the Bible, there are numerous assertions that God exists in a realm distinct from our physical world. In answer to Pilate's question about his kingship, Jesus clearly states: "My kingdom is not of this world. If My kingdom were of this world, My servants would be fighting so that I would not be handed over to the Jews; but as it is, My kingdom is not of this realm" (John 18:36). We also can infer through the creation accounts in Genesis 1 and in other biblical references that God created the heavens and the earth and, therefore, that he existed and continues to exist outside of the created universe. Connected with this assertion is the important conclusion that God has dominion over his created world, our universe. We have this statement in Colossians 1: "For by Him all things were created, both in the heavens and on earth, visible and invisible, whether thrones, or dominions, or rulers, or authorities—all things have been created through Him and for Him. He is before all things, and in Him all things hold together" (Colossians 1:16–17).

Many scientists have trouble with such a statement, because they would say that God's dominion over the natural world would make it difficult to understand the laws of nature. However, if we accept that God has established natural laws as patterns for the orderly functioning of the universe, the scientific method is never brought into conflict with God's superintending of creation at a very basic level. Deborah and Loren Haarsma, in their book *Origins: Christian Perspectives on Creation, Evolution, and Intelligent Design*, have described God's governance over the world through natural laws. They point out that while the physical world can be described by natural laws, "that does not make natural law identical to how God actually governs." They point out that it is more accurate to say that "God usually governs nature in regular patterns, and to the extent that we understand these patterns, we call them natural laws."[13] In this worldview, we can see that God's governance, or holding things together, is not in conflict with the carrying out of science and how we understand the natural laws that he has established. Moreover, this view points to God being everywhere present in the universe, upholding the natural laws, and maintaining the physical principles that we can know through the scientific method.

One might ask: "How could a book such as the Bible, written in ancient times, be taken seriously in our time with the advanced knowledge that we have of the universe?" Two points need to be made about this objection: First, the ancients (i.e., the Hebrew writers) were limited by what they knew of the physical world. Second, the writers of the early chapters of the Bible were more concerned with God's participation in man's redemptive history than they were about the details of how he had created the world. In the writing of the scriptures, the intent was not to give physical details of God's creation of the world or to investigate the physical world. We see in the early chapters of Genesis a brief poetic description of God's creation of the world, and then the prime focus is on man and his relationship to God. We see God revealing himself to man so that man can be redeemed from an inherent sinful nature and brought into a positive relationship with God.

The ancients, particularly the Hebrews of the second millennium BC, had a view of the physical world that was for the most part inherited from other cultures of the time (i.e., the Sumerian and Babylonian civilizations). The early Hebrew cosmological view of the world has been referred to by N. F. Gier as "the three-story universe" (see figure 3.2). This model envisions a solid dome (the firmament) suspended over the flat disk of the earth. Below the earth is the underworld (or Sheol), and all is supported by the pillars of earth, suspending the earth in the primeval ocean. The firmament served to fix the sun, moon, and stars. The solid dome of the firmament also separated the earth from the ocean of heaven (the waters above) and the chambers in heavens (the dwelling place of God).[14]

Gier has referenced from the evangelical Joseph Dillow that the Hebrew scriptures were written based on the doctrine of "sharable implications." This doctrine states, according to Gier, that "we cannot impute to the authors knowledge or experience which they could not possibly have had."[15] This is an important concept in biblical interpretation, as will be seen in later chapters, because it serves to explain why the biblical writers of the early chapters of Genesis expressed the creation and other stories as they did.

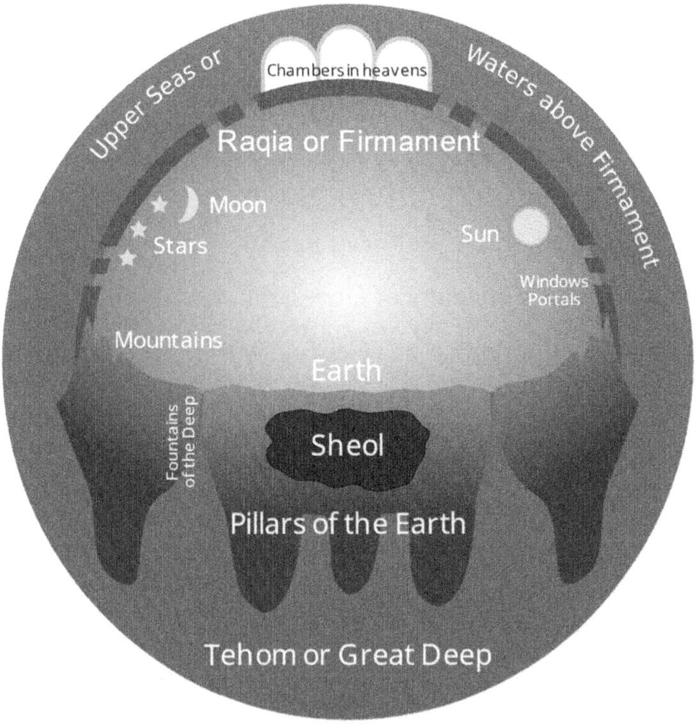

Figure 3.2. Early Hebrew conception of the universe (from commons.wikimedia.org) (adapted from R. V. Chamberlin, "The Early Hebrew Conception of the Universe," White and Blue 13, no. 11 [1909]: 84).

Accordingly, they recorded the accounts as they experienced them, and not according to scientific knowledge of the world that would be later revealed. Applying this concept of "sharable implications" to the Hebrew model of the three-story universe, we can conclude that it is likely that the Hebrews viewed the earth and heavens according to what they observed at the time. While this was a limited view, it did not impair their theological understanding of God or his redemptive plan for mankind.

This has great implications for many of the stories in the early chapters of Genesis. For example, the account of the flood of Noah (Genesis 7–8) states that as a result of the flood, "all creatures that moved on the earth perished: birds, livestock, animals, and every swarming thing that swarms upon the earth, and all mankind; all . . . died" (Genesis 7:21–22). According to "sharable implications," we can interpret this verse to indicate that all the people and animals that they could see were obliterated by the flood. This does not necessarily mean that this was a universal flood encompassing the entire earth. Interestingly, in reading the biblical text, we find very little to indicate a flood encompassing the entire earth. It is only indicated that all the known earth was covered with water, and all known life was obliterated. Another example of "sharable implications" is the apparent age of the biblical patriarchs in Genesis 1–11. From Adam to Noah, the life spans of all the patriarchs except Enoch are given as ranging from 895 years to 969 years (see Genesis 5:1–32; 9:28–29).[16] Abraham, who lived at least eleven generations after Noah, is noted to have lived 175 years (Genesis 25:7). At the end of Genesis we see, a number of generations later, that Joseph is recorded to have a life span of 110 years (Genesis 50:26). A number of hypotheses have been put forth to explain why the initial patriarchs apparently lived so long and why their life spans declined over time. The doctrine of "sharable implications" might suggest that the early patriarchs did not have good timekeeping skills. While this is only speculation, it is likely that the earlier generations from Adam through Abraham, living in Mesopotamia, may not have had good timekeeping skills. Thus, they may have greatly over-estimated the life spans of the earlier family members. It is likely that as the later generations moved to Egypt, they benefited from Egyptian astronomical and timekeeping skills, which would have given them life spans more in the range of what we experience in current times. The major take-home message here is that the biblical genealogies and events recorded in the early chapters of the book of Genesis may be subject to "sharable implications" that limit our ability to interpret the texts literally. However, this does not limit the ability of the texts to speak faithfully in terms of their theological messages.

A major advance from the three-story universe cosmology came about during the Alexandrian age of the classical Greek period (ca. 300 BC to AD 100) under the Alexandrian family of Ptolemy, established in Egypt after the death of Alexander the Great. Ptolemy I Soter (367–282 BC), who was the first pharaoh of the Ptolemaic Kingdom and studied under Aristotle, contributed to advancement in cosmology by founding the Museum of Alexandria in Egypt.[17] Modeled after the Greek Lyceum, the museum was a research and teaching institute that included a zoo, botanical gardens, an astronomical observatory, dissecting rooms, and a library with about half a million scrolls. Several centuries later, Claudius Ptolemy (ca. AD 85–165; hereafter "Ptolemy") built on the cosmology he inherited. Ptolemy, a mathematician, astronomer, and geographer in Alexandria, developed geometric models of the motions of stars and planets derived from astronomical observations over more than eight hundred years. Ptolemy developed an earth-centric model—often called a geocentric model—of the universe, with all other celestial bodies revolving around the earth in concentric circles or spheres. This model became universally accepted in the Western world (see *Ptolemy's Catalogue of Stars: A Revision of the "Almagest"*[18]) and remained dominant until the sixteenth century, when heliocentric models were developed.

Nicolas Copernicus (1473–1543), a European astronomer from Poland, developed a heliocentric model of the solar system where all planets revolved around the sun in concentric circles. Copernicus suggested that this model fit better with the observed motion of the planets than the earth-centric Ptolemaic system. Copernicus based his conclusions on two propositions:[19]

> *(1) They must "save the appearances" (i.e., agree with observed motion of the planets).*
>
> *(2) They must not contradict Pythagorean mathematics of the motion of the planets.*

Copernicus published his work on the heliocentric universe, *On the Revolutions of the Celestial Orbs*, in 1543, shortly before his death.[20]

A later and more correct model of the solar system was proposed by Johannes Kepler (1571–1630), who proposed that all planets traveled around the sun in elliptical patterns, with the Sun at one focal point of the ellipse. Kepler based his conclusions on the observations of Tycho Brahe (1546–1601), a Danish nobleman who assumed a modified geocentric solar system but had made observations superior to those previously used to develop the Ptolemaic model. Kepler described the motion of the planets around the sun by three laws:[21]

> *(1) Each planet follows a path described by an ellipse, with the sun at one focal point.*
>
> *(2) The line drawn from the sun to the planet sweeps out equal areas at equal times.*
>
> *(3) The squares of the times the planets require to complete their orbits are proportional to the cubes of their respective mean distances from the sun.*

This cosmological model of Kepler was later confirmed by Galileo Galilei (1564–1642) in his fundamental works based on his own telescopic observations of Jupiter and other planets.[22]

The development of astronomy through the seventeenth century not only helped to establish the scientific method, leading to the age of the scientific revolution; it also led to an expanded understanding of the cosmos. However, it did not push God out of the picture. In a nod to the creator of the universe, Galileo wrote in 1610 following his initial telescopic observations of the heavens: "I render infinite thanks to God for being so kind as to make me alone the first observer of marvels kept hidden in obscurity for all previous centuries."[23] We can gather from Galileo's statement that he regarded the divine attributes to be as real as the physical qualities that he so carefully observed. He saw no conflict in giving thanks to God for the opportunity to carry out his work as a scientist.

∞

Why is it that so much science is investigated, and yet many scientists fail to see the hand of God? The answer to this may be that by the nature of science and the scientific method, it is all done close up to nature itself. Scientists often study one enzyme, one type of cell, or one organism for their entire careers. They publish reams of papers on that one subject. This is certainly necessary and commendable. The work needs to be done under controlled conditions without any sense of bias to supernatural intervention. We do not allow the divine to enter into such work. We only allow for the regular and orderly patterns that are seen in controlled scientific observation. It is important that we draw our conclusions and formulate our theories based on logical and sound data, the essence of the scientific method. Many times, we see the grandeur of God working through the physical universe. However, for the most part we do not see the hand of God until many works of science can be synthesized and viewed from a distance. When Michelangelo painted on the plaster ceiling of the Sistine Chapel, he was working close-up, painting minute details like hairs and facial features. It is not until we stand at a distance that we can see the hand of God reaching out in *The Creation of Adam* (see figure 3.3).

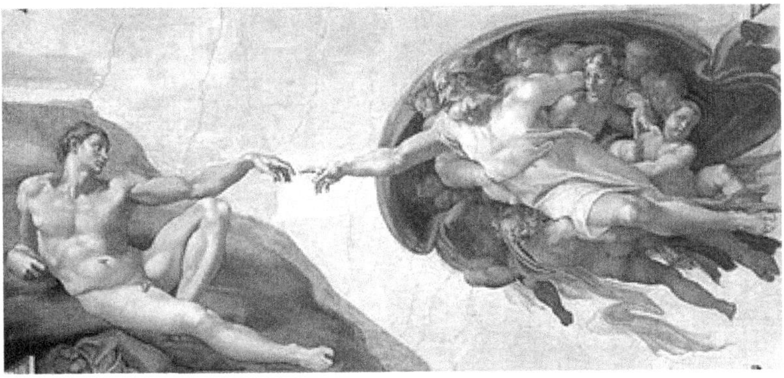

Figure 3.3. Michelangelo, The Creation of Adam (from Wikimedia.org).

Likewise, the conclusions of Galileo were the result of more than fifteen hundred years of observation by a multitude of astronomers working on isolated details. It is only when we stand back and view these many findings in perspective that we see the work of God in our physical cosmos. In the coming chapters, I will detail areas of investigation in the life sciences from work by scientists over the last approximately one and a half centuries in the areas of structural biology, molecular biology, enzymology, and evolution. It is minute and microscopic work looking deep into the structure and function of living systems. These works can be compared to individual pieces in a giant jigsaw puzzle. As with the individual pieces of the jigsaw puzzle, looking at the individual scientific findings, we cannot see the overall pattern. However, if we stand back and view these findings in perspective, we will see the truly magnificent work of God.

Chapter 4:

Exploration of the Physical Realm through the Study of Structural Biology

In the life sciences, biologists investigate structure and function both at the molecular level and at the cellular or organismal level. It is an underlying principle that properties observed at the molecular level have direct consequences for observations made at the cellular or organismal level. This principle has been a driving force in the study of structural biology—the determining of three-dimensional structures of biomolecules. No area of biology has been untouched by structural biology. The determining of three-dimensional molecular structures of proteins, nucleic acids, and other cellular components has been central to the understanding of almost all biological processes. By determining the 3-D structures of biomolecules at a high resolution, we not only see the molecular shape, but we also can postulate how these molecules function and how they interact with living systems.

Just as the heavens reflect the glory of God, as indicated in the quote from Psalm 19 at the beginning of chapter 3, so these 3-D molecular structures also reflect the glory of God in their functioning at all levels of living systems.

I first encountered structural biology in 1972 as a freshman at Johns Hopkins University in Baltimore. In a biophysics seminar, we were introduced to the 3-D structure of proteins through a book entitled *The Structure and Action of Proteins*, by Dickerson and Geis.[24] In his introduction to the book, Max Perutz stated that "the recent development of chemical and x-ray crystallographic methods for solving the structure of enzymes has been one of the triumphs of molecular biology."[25] Based on the subsequent influence of structural biology on biology as a whole, this is indeed an understatement. When the book was published (1969), only about a dozen enzymes and proteins had been subjected to X-ray crystallographic analysis to yield high-resolution 3-D structures. To date, more than 170,000 protein and nucleic acid 3-D structures have been solved, including a small portion of structures solved by a technique called nuclear magnetic resonance (NMR).[26] New structures are currently being solved at a rate of approximately 10,000 per year.

In our biophysics course, we were able to view projected 3-D structures of hemoglobin and myoglobin using Polaroid glasses similar to those used at 3-D movies. The protein molecules jumped off the screen. Folded proteins with a helices and b sheets were seen in a depth not previously viewed in two dimensions. Scientists viewing these molecules in 3-D could better interpret the interactions of different protein chains through specific amino acid residues via hydrogen bonding, disulfide bonds, and other covalent interactions. Residues could also be seen to interact via hydrophobic interactions. Space-filling CPK models of proteins allowed the visualization of crevices and pockets that were shaped to allow specific binding of substrate molecules. However, these 3-D structures were viewed (through slide projections, physical models, or computer views), these models allowed the scientists to visualize a close-up, high-definition view of the geometric space that these molecules occupied. The models also allowed the scientists to propose mechanisms of how these molecules carried out their complex chemical reactions and molecular functions. As a student, I had my view of the physical world expanded so that I could now see biomolecules as dynamic structures that functioned in an orderly way to carry out the features of living systems. Whether studying molecular, cellular, or physiological functions, scientists now understood that almost all biological phenomena were to be explained in terms of their structure-based biomolecular components.

The origins of structural biology can be traced back a mere fifty years before the time that crystallographers began to unravel the 3-D structures of proteins in the 1950s through methods developed by physicists. Noted physicist Michio Kaku, in his video "The Universe in a Nutshell," has commented about our modern technology that "almost everything you see in your living room, almost everything you see in a modern hospital can be traced to a physicist."[27] In a similar way, the entire development of structural biology can be traced back to a physicist: William Lawrence Bragg. Born in Australia in 1890 to noted physicist William

Henry Bragg,[28] as a child, William Lawrence fell off his tricycle and broke his arm. William Henry had studied the work of Wilhelm Conrad Röntgen on the discovery of X-rays.[29] He had been developing his own X-ray apparatus, and he used this device to view his son's broken arm to confirm that the bone was in fact broken, thus becoming the first in Australia to use X-rays for medical diagnosis.[30] William Lawrence pursued an education in mathematics, chemistry, and physics at Adelaide University in Australia.[31] His father accepted a position at the University of Leeds in England as a physics professor in 1908. When the family moved to England, William Lawrence enrolled as an undergraduate at Trinity College, Cambridge, where he majored in physics and graduated with first class honors in 1914.

William Lawrence Bragg then pursued graduate work in physics at Trinity College. During that time, William Henry Bragg's continuing investigation into X-rays led him to postulate that X-rays, like other forms of radiation, were particulate in nature. However, the son, William Lawrence, had a hunch based on the work of Max von Laue that X-rays were in fact wave-like in nature. Von Laue had irradiated a crystal with X-rays and observed a diffraction pattern of spots on a photographic plate, indicating that X-rays were electromagnetic waves, like visible light.[32] From this work, William Lawrence Bragg and his father developed an X-ray spectrometer and carried out work on irradiation of crystals by X-ray beams at different angles of incidence that enabled them to determine the dimensions of the spacing of the atoms within the crystal through what was to become known as the Bragg equation. As Bragg described it,

> *The X-ray spectrometer opened up a new world. It proved to be a far more powerful method of analyzing crystal structure than the Laue photographs which I had used. One could examine the various faces of a crystal in succession, and by noting the angles at which and the intensity with which they reflected the X-rays, one could deduce the way in which the atoms were arranged in sheets parallel to these faces. The intersections of these sheets pinned down the position of the atoms in space. . . . It was like discovering an alluvial gold field with nuggets lying around waiting to be picked up. . . . It was a glorious time when we worked far into every night with new worlds unfolding before us in the silent laboratory.*[33]

For this work, father and son received the Nobel Prize in physics in 1915. William Lawrence Bragg, who was twenty-five years old, was the youngest to receive the Nobel Prize at that time.[34]

Later, as a professor at the University of Manchester, William Lawrence Bragg (hereafter "Bragg") continued his work in X-ray diffraction of more complicated, inorganic crystals, such as silicates. He found that even in these more complicated crystals, the positions of the different atoms could be determined through the measuring of the intensity of the diffraction spots at different angles of incidence and the use of mathematical relationships known as Fourier transforms.[35] Determination of 3-D structures of the more complicated crystals was critical in the later determinations of 3-D structures of biological proteins whose molecules had thousands of atoms.

In 1938, Bragg was appointed professor and head of the Cavendish Labs at the University of Cambridge. The Cavendish Labs carried out a wide variety of cutting-edge physics research. Bragg headed the X-ray crystallography group. Among the group was an immigrant from Austria named Max Perutz.[36] Perutz had done some preliminary X-ray diffractions of the protein hemoglobin at the suggestion of J. D. Bernal, a crystallographer at Birbeck College of the University of London and at the Cavendish Labs. Perutz convinced Bragg that it would be possible to determine the 3-D structure of hemoglobin from X-ray diffraction patterns. Thus started a more-than-ten-year effort, headed by Perutz, to determine the 3-D structure of hemoglobin (1947–59).

At the Cavendish Labs, Bragg was as much an able administrator as he was a scientist. He organized the labs into small groups of six to twelve scientists and a few assistants and instrument specialists.[37] While continuing with his own work on diffraction of X-rays by metals, he encouraged others in his group to pursue work on complex biomolecules such as nucleic acids and proteins. While he managed these projects, he always gave full credit to those who performed the work. Bragg also deeply appreciated the intricacies of the protein structures that his lab had determined. His daughter, Patience, related an experience that illustrates this appreciation and hints that he may have seen spiritual significance in natural phenomena he studied. She and her father were traveling by airplane to Berlin, where he would give a lecture on the protein lysozyme. Patience had a 3-D model of the protein lysozyme on her lap as they traveled (see figure 4.1). Bragg commented about the structure, "It is such a pleasing solution and so beautiful. It must be right." At the airport in Berlin, a large crowd met them and presented a bouquet of flowers to Patience. In her narrative of the experience, Patience commented that Bragg was not a religious man in the traditional sense. Bragg regarded himself as a "blue sky" worshiper who chose to work in the garden when the rest of the family attended church. With respect to theology, Patience wrote, Bragg felt that "the evidence of God was in the visual world and that whatever the arrangements for the afterlife turned out to be, he was convinced that they would be satisfactory."[38] We can infer that Bragg saw evidence for God as he looked at these protein structures. We can also make our own conclusions about God working in these intricate structures, as much as we do when we see distant galaxies in the far reaches of the universe.

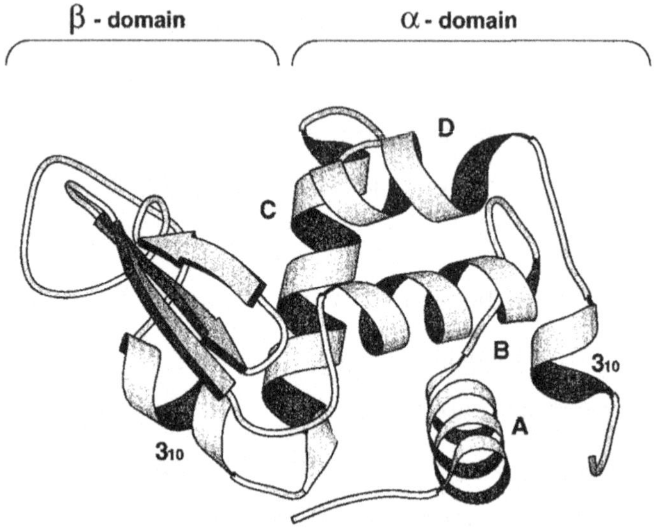

Figure 4.1. Chicken egg white lysozyme 3-D structure (from B. Gilquin, C. Guilbert, and D. Perahia, "Unfolding of Hen Egg Lysozyme by Molecular Dynamics Simulations at 300k: Insight into the Role of the Interdomain Interface," PROTEINS: Structure, Function and Bioinformatics 41 [2000]: 58–74, with permission).

Max Perutz was short in stature but a giant in the field of protein crystallography. Born in Austria, he received an undergraduate degree from the University of Vienna in 1936. Because of the German invasion into Austria, Perutz and his parents fled first to Switzerland, and then to England.[39] After making numerous connections at the University of Cambridge, Perutz became a research student under J. D. Bernal at the

Cavendish Labs.[40] Perutz was encouraged by Bernal to study the 3-D structure of proteins as determined by diffraction X-ray. In 1939, Perutz received a grant from the Rockefeller Foundation to establish work in the study of the crystallographic study of hemoglobin. Perutz eventually earned his PhD in 1939 for his study of the X-ray diffraction of horse hemoglobin under Bragg.[41] During World War II, Perutz was interned briefly by the British because of his Austrian origin, and then served in the British Army as an engineer for various defense-related projects. After the war, he returned to the Cavendish Labs to continue his work on hemoglobin structure. In 1947, he received funding from the Medical Research Council (MRC) and founded the Molecular Biology Unit at the Cavendish Labs.[42] Perutz attracted a number of talented students to study protein and nucleic acid structures. These students included John Kendrew, James Watson, and Francis Crick. Perutz made his first major contribution to X-ray crystallography of proteins in 1953 when he solved the phase problem.[43] In order to determine the phases, Perutz compared X-ray diffraction patterns of protein crystals with heavy atoms (silver or uranium) and X-ray diffraction patterns of protein crystals without heavy metals, a process known as multiple isomorphous replacement. While the method was successful, it took six years to complete the solution of the hemoglobin 3-D structure at high resolution. Perutz's contribution was notable because his method of isomorphous replacement became the standard method for X-ray 3-D structure determination. Advanced computer methods, developed since then, have greatly shortened the time required for determining electron density maps.

Perutz is also known for determining the comparative structures of oxy- and deoxy- forms of hemoglobin. From these forms, he was able to determine the conformational changes resulting from oxygen loading of hemoglobin. Perutz was able to explain how pH controlled the loading of oxygen onto hemoglobin in the lungs and the unloading of oxygen in tissues where it was transferred from hemoglobin to myoglobin.[44] This correlation of conformational structural change was a model for determination of structure/function for a variety of enzymes whose structures were being solved during the 1970s. Perutz showed that protein 3-D structures were not static, but were dynamically changing during enzymatic reactions and allosteric binding of regulatory molecules.[45]

The road to the first high-resolution protein 3-D structure by X-ray crystallography was a more-than-ten-year journey. Most of the work to determine the 3-D structure of myoglobin was conducted at the Cavendish Labs MRC Molecular Biology Unit under the direction of John Kendrew. In the early 1950s, many of Kendrew's colleagues considered the determination of protein 3-D structures from X-ray data to be far too complicated to be achievable. Michael Rossmann, a coworker of Perutz, commented: "The likelihood of ever solving the 3-dimensional structure of a protein seemed exceedingly small even in the 1950s. There were too many data to be collected, no reasonable way of solving the phase problem, and no way of computing results to make an interpretation of the collected data. But even more daunting was the nagging question of whether each protein molecule had a unique structure."[46]

John Kendrew could be described of a scientist among scientists. He was born into a family of academicians in England in 1917. His father was a respected climatologist, his mother an Italian art historian. When he was four, his father and mother separated, and from then on, he was raised by his father, Wilfrid Kendrew.[47] In spite of the family challenges, Kendrew received a liberal education in English private schools. At the age of fifteen, he determined that he wanted a career in science. As a high school student at the Clifton School, Kendrew was taught by a strict teacher who required that he learn organic chemistry from a German textbook. He taught himself German by studying a German Bible that he found in the chapel and was soon able to read the German chemistry textbook. At Clifton, his teachers drilled him on the ability to do exacting mathematical calculations without error. Over his years from high school through graduate school, he kept detailed notebooks of lectures, carefully organized by subject and course designation. He kept these notes up until his death in 1997. After graduating from Trinity College in 1939, Kendrew entered military service in operations research as England entered World War II. During this service, which took him to many places around the British Empire, Kendrew worked with J.D. Bernal. Bernal encouraged Kendrew to pursue X-ray crystallography of proteins and was instrumental in supporting Kendrew's application for a graduate student appointment to the Cavendish Labs under William Lawrence Bragg.[48] Kendrew initially worked under Max Perutz on the hemoglobin X-ray structure. Perutz had been working on hemoglobin since 1936 and had made limited progress toward the 3-D structure. Though Kendrew lacked formal training in X-ray crystallography when he joined Perutz, he learned the subject in about two weeks and was very highly regarded by Perutz for his scientific ability.[49] Kendrew received his PhD in 1949 in X-ray diffraction comparisons of adult and fetal hemoglobin.

Kendrew felt that a less complex protein would be easier to solve, so after graduation, as he continued his work at the Cavendish Labs, he concentrated on the structure of myoglobin, the oxygen carrier in muscle tissues. Myoglobin was later shown to have a structure similar to that of the individual subunits of the hemoglobin tetramer, including the oxygen-binding heme component. In the early 1950s, Kendrew concentrated on two particularly daunting problems that would have to be addressed in order to determine the myoglobin structure: (1) obtaining quality crystals of myoglobin and (2) doing complex calculations of the thousands of X-ray diffraction spots. Kendrew solved the first problem by choosing sperm whale myoglobin as the source of protein. Sperm whales, because of their physiology as underwater mammals, have a large amount of myoglobin in their muscles. This source yielded large, high-quality crystals, and these crystals gave reproducible high-quality X-ray diffraction patterns.[50] The second problem, computation of the data, was perhaps more difficult, because digital computers in the early 1950s were slow and primitive in terms of programming capacity. However, by using an available computer at the Cavendish Labs, the EDSAC-I, rather than a punch card machine, Kendrew was able to greatly speed up calculations, so that instead of the usual six months they took thirty to forty hours.[51] In 1958 the EDSAC-I was replaced by the much faster EDSAC-II.

Another problem that stood in the way of X-ray structure determination was known as the phase problem. As indicated above, this problem was eventually solved by Max Perutz in 1953. X-ray diffraction patterns yielded the intensity and position of the diffracted spot and allowed the calculation of the basic repeat structure of the molecule within the crystal. However, to calculate electron densities within the repeat unit, another piece of data was needed—the phase of the diffracted waves. Perutz conjectured based on previous work by the chemist Johannes M. Bijvoet that heavy metals could be complexed with the molecule of interest and that comparison of the heavy metal–complexed diffraction pattern with the noncomplexed diffraction pattern would yield information on the phase of the waves. This was a process that became known as multiple isomorphous replacement. When silver or mercury atoms were complexed with hemoglobin, the diffracted spots appeared in exactly the same positions as the noncomplexed crystals. This indicated that the complexed metals had not changed the protein 3-D structure. However, the intensities of the spots were slightly altered, and this change allowed the calculation of phases for each diffracted spot.[52] This new method became a standard for X-ray crystallographers to calculate 3-D electron density maps for molecular structure determination. John Kendrew then applied this new methodology to the determination of the 3-D X-ray structure of sperm whale myoglobin. After more than four years of work, Kendrew and his colleagues published the 6Å resolution 3-D structure of sperm whale myoglobin in 1957. After collecting further data, he published the 2Å resolution structure in 1959 and the 1.4Å resolution structure in 1965 (see figure 4.2). It was only in this highest-resolution structure that amino acid side chain positions could be accurately determined. After he determined the first low-resolution 3-D structure of sperm whale myoglobin, Kendrew made the following comment in a 1961 *Scientific American* article:

> *When the early explorers of America made their first landfall, they had the unforgettable experience of glimpsing a New World that no European had seen before them. Moments such as this—first visions of new worlds—are one of the main attractions of exploration. From time to time scientists are privileged to share excitements of the same kind. Such a moment arrived for my colleagues and me one Sunday morning in 1957, when we looked at something no one before us had seen: a 3-dimensional picture of a protein molecule in all its complexity.*[53]

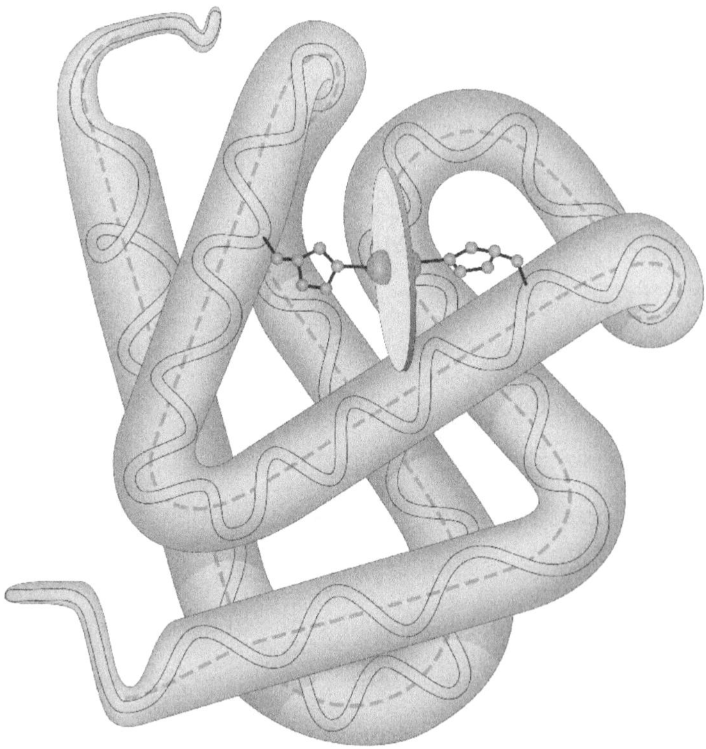

Figure 4.2. Human myoglobin 3-D structure (courtesy of A. Griskeviciene, Shutterstock, https://www.shutterstock.com/image-illustration/structure-human-myoglobin-molecule-1179943432 [royalty-free permission]).

These studies on sperm whale myoglobin set the stage for Max Perutz's determination of the 3-D X-ray structure of hemoglobin in 1960 and for numerous determinations of enzyme 3-D structures carried out at the MRC Molecular Biology Unit through the 1970s. In 1962 Kendrew and Perutz shared the Nobel Prize in Chemistry for their work in X-ray 3-D protein structure determination.

Most of us, with ready access to these 3-D structures, have come to accept them as part of our scientific or physical reality. When I first saw some of these structures back in the 1970s, I was impressed by the beauty, depth, and geometric elegance of these structures. However, it is worth asking whether these models reflect merely a physical reality, or whether they might also reflect a spiritual reality of the God who might have created them. Both Perutz and Kendrew were atheists by religious profession.[54] Thus, they probably didn't see things from a spiritual perspective, even though they, like Bragg, probably appreciated the chemical geometry and the molecular workings of these structures. In his introduction to *The Structure and Action of Proteins*, Perutz states: "So far no novel or sophisticated forces have had to be invoked to explain their action. They can be understood in terms of simple electrostatics. Enzymes recognize their specific reactants by having a site that fits them as a lock fits a key, only more so: the lock often carries electric charges or dipoles exactly compensating those of the reactants."[55] I would agree with this in that many of these phenomena of enzyme action can be explained by electrostatics and by hydrophobic and hydrophilic interactions. However, even Perutz was aware of the dynamic actions of enzymes and also of hemoglobin in its ability to rapidly change conformations from oxy- to deoxy- forms. What we see in an X-ray crystallographic structure is a mere snapshot of the enzyme as it functions dynamically in the cell. In a functioning enzyme, substrates are converted to product at a rate of thousands per second, something that we cannot possibly imagine. In the 1930s, Harold Edgerton, a professor of electrical engineering at the Massachusetts Institute of Technology, developed a technique to photograph fast-moving objects such as a bullet fired from a gun or a drop of milk impacting a bowl of liquid.[56] By using an electronic strobe and a motion-picture camera, Edgerton was able to capture the motion of fast-moving objects in a way that we could visualize them (see figure 4.3).

Figure 4.3. Strobe photography by Dr. Harold Edgerton captures a diver in motion.[57]
© 2010 MIT. Courtesy of MIT Museum. With permission.

If we could capture the dynamic function of enzymes with an imaginary strobe, we might see far more than we do with the static snap shots that X-ray crystallography allows. We do not have to invoke anything other than physical forces that bring this about. However, we might see in such a hypothetical live-action

3-D view something more of God's creation, just as Galileo saw in the motion of the planets. If we acknowledge God's ability to govern the laws of this 3-D world that we live in, we do not downgrade the precision to which these laws are governed. We do not limit the ability of the scientific method to investigate these laws. On the contrary, we merely enlarge our worldview to attribute these wide-ranging laws of nature to a transcendent God who has dominion over the earth as we observe it.

Chapter 5:

The Birth of Molecular Biology and Exploration of the Genome

Molecular biology originated with the determination of the 3-D, double-helical structure of DNA by Watson and Crick in 1953. The area of work quickly shifted from structural considerations to the mechanism by which the genetic material coded for proteins and to the actual sequences of DNA that coded for those proteins. Eventually this would lead to the sequencing of entire genomes of organisms. In this section, we will consider the solving of the double helix in 1953 and the sequencing of the human genome in 2003—a period of almost exactly fifty years—as bookends between which molecular biology flourished in the latter part of the twentieth century. We will also consider events in the 1940s that led up to the solving of the double helix. We will examine the work that followed the sequencing of the human genome as an indication of where genomics may go in the twenty-first century. Lastly, we will consider what are the physical and spiritual implications of knowing our genetic makeup. As many have speculated, our increased understanding of genetic alterations that affect certain diseases will likely revolutionize human medicine. However, knowing our genome in detail also has great implications for how we view ourselves as a species, how we relate to other life forms, and what we consider to be our place in the universe. If God is indeed our creator, what does the human genome say about how that creative activity occurred? If we were created in God's image, do these sequences say something about the nature of God himself? It is again important to have an overview of how this vast amount of information in the genome affects our picture of our place in the universe, as well as how it may change our physical lives in the years to come.

When James Watson and Francis Crick signed on at the MRC Unit of the Cavendish Labs, they had two things in common. They were both convinced that DNA was the genetic material, and they both believed that X-ray crystallography could reveal the 3-D structure of DNA. Watson's account of the DNA structure solution in *The Double Helix* is a story of how two maverick biologists spent just a few years pouring over the data of other scientists, carrying out discussions over beer in English pubs, and building 3-D models to come up with a structure solution that would literally change the face of the biological sciences.[58]

Both Watson and Crick were aware of work done by Oswald Avery in 1944 indicating that DNA was the genetic material. Avery's experiment indicated that DNA was the "transforming principle" allowing the induction of a lethal phenotype in Streptococcus pneumococci Type III bacteria with which mice were infected. Avery determined the identity of this "transforming principle" by separating protein components from the DNA and digesting polysaccharide components enzymatically.[59] Protein components in this transforming preparation were shown to be at most 0.02%. Avery found that proteases had no effect on the transforming ability of this preparation, but a "deoxyribonuclease" did destroy the transforming ability of the preparation. This experiment indicated that DNA and not protein was the genetic material required for transformation of the pneumococcus bacteria. DNA was confirmed as the genetic material in 1952 by Alfred Hershey and Martha Chase, who showed that transformation of bacteria by bacteriophages resulted from injection of the DNA and not the protein.[60]

The earliest X-ray diffraction patterns from Maurice Wilkins' lab were suggestive of but did not conclusively show the helical structure of DNA. In fact, Rosalind Franklin, who carried out the initial X-ray diffraction of DNA, very much opposed the idea that the patterns indicated a helical structure. Watson, however, based on his work with tobacco mosaic virus, and being aware of Linus Pauling's determination of the protein α-helix, strongly favored the idea of a DNA helical structure. Another piece of critical data was provided by Erwin Chargaff of Columbia University, who found that quantitatively the number of adenine bases in DNA was nearly equal to the number of thymine bases, and the number of guanine bases was nearly equal to the number of cytosine bases.[61] It was unclear what the significance of this data was, but the two detectives from the Cavendish Labs would eventually piece this data together in their 3-D model of DNA. Another critical piece of data came from a new crystal form of DNA referred to as the DNA B form. Rosalind Franklin had crystallized DNA with a higher water content. These crystals gave an X-ray diffraction pattern different from the previously crystallized A form. The X-ray diffractions of the DNA B form suggested a more definite helical structure. Furthermore, the patterns suggested a packing of the bases with a 3.4Å repeat and a helix repeat of 34Å (see figure 5.1).[62]

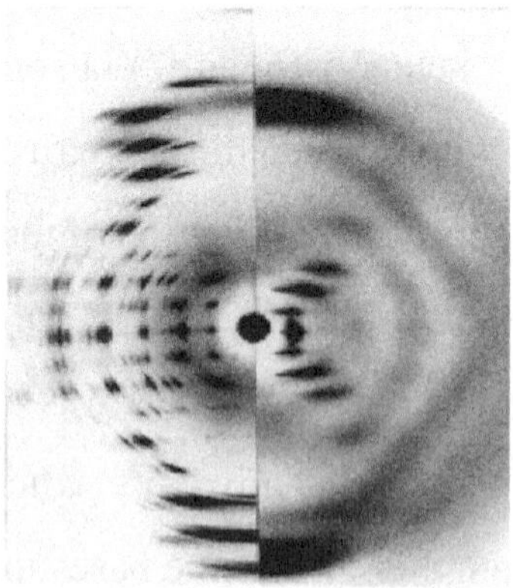

Figure 5.1. X-ray diffraction patterns of DNA A- and B- crystals (see I. C. Baianu et al., "File:ABDNAxrgpj.jpg," Wikimedia Commons, last edited May 14, 2021, https://commons.wikimedia.org/wiki/File:ABDNAxrgpj.jpg).

The dimensions were sufficient for Watson to start building models with the assumption that two phosphodiester backbones formed the external structure of DNA. Watson's first model assumed that base pairs of the opposing strands were held together by hydrogen bonds between like bases in the core of the structure (i.e., A-A, G-G, C-C, T-T). This model was quickly shot down by a crystallographer in the lab, Jerry Donahue, who pointed out that Watson was assuming the enol forms of the bases. The predominant keto forms of the bases could not associate as Watson had hypothesized.[63] Watson was playing around with cardboard models of the keto forms and found surprisingly that A-T and G-C keto forms of the bases could bind together by hydrogen bonds in a geometry that fit perfectly within the double-helix backbone. It then dawned upon him that such associations would satisfy the Chargaff Rules, which indicated equivalence of A-T and G-C numbers of base pairs.[64] After he and Crick built their model with A-T and G-C base pairs on the inside and two ribose-phosphodiester backbones in opposing directions, they were convinced that this double-helical model was in agreement with the X-ray diffraction pattern of Rosalind

Franklin. Not only did this model agree with all the data, but it also suggested a mechanism by which DNA strands could be replicated prior to cell division (see figure 5.2 and figure 5.3).

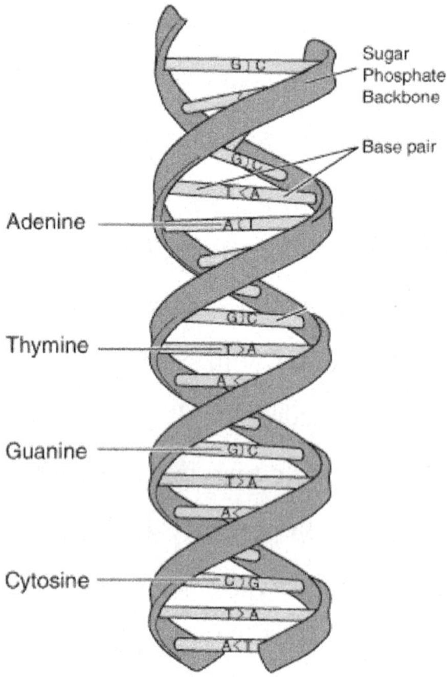

Figure 5.2. DNA 3-D double-helical structure with sugar phosphate backbones on outside and AT and GC base pairs on the inside (see "DNA structure and bases.png," Wikimedia Commons, last edited [01-12-2006], [https://commons.wikimedia.org/wiki/File:DNA_structure_and_bases.png], with modification).

When James Watson and Francis Crick solved the 3-D structure of DNA in 1953, they wrote a short paper to be published in the journal *Nature*, describing their findings.[65] They had Watson's sister, Elizabeth, type up the final manuscript. There were no word processors in those days, only manual typewriters. According to Watson, in *The Double Helix*, they told Elizabeth "that she was participating in perhaps the most famous event in biology since Darwin's book."[66] If we were told in 1950 that a postdoc from the U.S. would come over to Great Britain and team up with a British doctoral student to write the most significant paper of the twentieth century, we would have said that this is a fantasy. In fact, the solving of the double-helix 3-D structure required the input of many scientists from Bragg's lab at the MRC Cavendish Labs (Francis Crick, James Watson, Max Perutz, John Kendrew, and others) as well as crystallographers at the Kings College in London (Maurice Wilkins and Rosalind Franklin). At the time, there was also major competition from Linus Pauling at the California Institute of Technology in the United States to solve the DNA 3-D structure. Viewed from the perspective of more than sixty years of biological research, the 1953 paper by Watson and Crick stands as a turning point in the history of how we view biology, and most agree that the paper has lived up to Watson's original assessment.

The period immediately following the revealing of the structure of DNA led to a detailed mechanism by which DNA replicated and also coded for the expression of proteins. In 1958, Matthew Meselson and Frank Stahl carried out an experiment that confirmed the semiconservative replication of DNA as proposed by Watson and Crick. Initially, DNA from bacteria grown in ^{14}N-containing medium was shown to migrate at a low density on CsCl gradients in an analytical ultracentrifuge. By carrying out multiple replications of bacteria in ^{15}N-containing medium, DNA from these bacteria sedimented at a higher density on CsCl salt gradient sedimentation. After a single round of replication in ^{14}N-containing medium, the DNAs now showed migration at a medium density on CsCl gradients. Further replications in ^{14}N-containing medium gave densities of DNA at predominantly the lower density. This confirmed that of the possible models, it was most likely that the DNA complementary templates were replicated by separation of the opposing strands and synthesis of new complementary strands to make up the replicated double-helical molecules (see figure 5.3). This mechanism is commonly referred to as semiconservative replication.

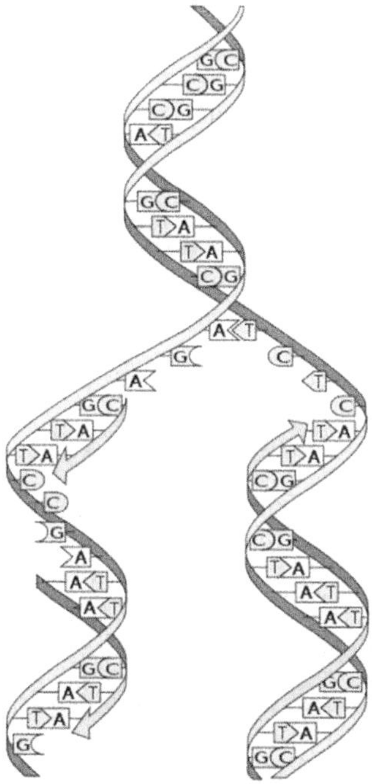

Figure 5.3. DNA double helix and semi-conservative replication (see Madeleine Price Ball "Unwound DNA Duplex.png," Wikimedia Commons, last edited [01-23-2013], [https://commons.wikimedia.org/wiki/File:Unwound_DNA_Duplex.png], with modification).

Another major question concerning the specification of proteins by DNA sequences was how the specific protein sequences were coded by the specific DNA base sequences. In 1957, Francis Crick had proposed the "Central Dogma" by which DNA sequences coded for a complementary RNA sequence, and these RNA sequences then coded for specific protein sequences. James Watson had also hypothesized a similar hierarchy of coding. Later on, in 1961, François Jacob and Jacques Monod proposed that a messenger RNA complementary to the DNA base pair sequence coded for the protein sequences. They also proposed that groups of proteins (operons) could be controlled by a regulatory sequence that acted upstream of the protein-coding sequences. During that time, Francis Crick and Sydney Brenner proposed that the coding to messenger RNA and amino acid residues was a triplet nucleoside code. Soon after this, Marshall Nirenberg and Heinrich Matthaei determined the specific nucleoside triplet codes that coded for the specific amino acids in proteins.[67] In 1956 it was shown that this coding was translated to amino acids by specific transfer RNA adapter molecules that associated a particular amino acid with each triplet code as specified by the messenger RNA sequence.[68] It was later shown that this process was mediated by the binding of messenger RNA in cooperation with the amino acid–bearing adapter molecule transfer RNAs at specific sites on the ribosome. By the middle 1970s the picture was more or less completed to describe the mechanism by which proteins were coded by DNA, transcribed into messenger RNAs, and translated into proteins by ribosomes (see figure 5.4). This "Central Dogma" of molecular biology has more or less stood the test of time, although some new details have emerged.

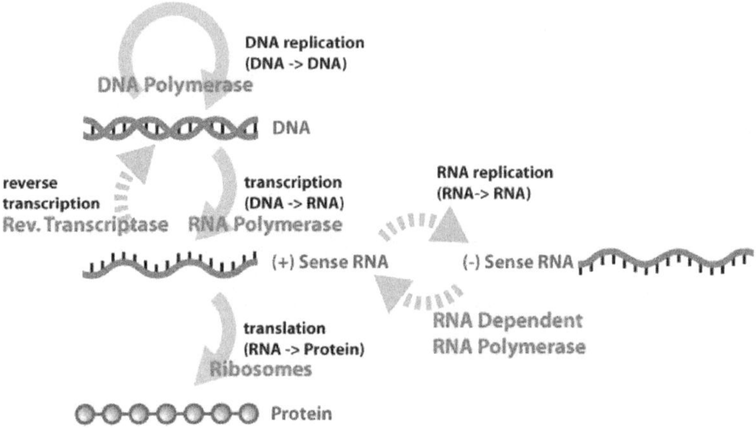

Figure 5.4. Central dogma for coding of proteins from DNA sequences via an intermediary messenger RNA transcribed on ribosomes (see Daniel Horspool "Extended Central Dogma with Enzymes.jpg," Wikimedia Commons, last edited [11-28-2008], [https://commons.wikimedia.org/wiki/File:Extended_Central_Dogma_with_Enzymes.jpg]).

In the early 1970s, there was a shift in molecular biology to the study of specific nucleotide sequences that coded for the proteins. Through the study of DNA replication in bacteria, techniques were developed for the propagation of mammalian gene sequences in bacterial plasmids. This so-called recombinant DNA technology was essential to the study of the mammalian DNA sequences. Restriction endonucleases, enzymes able to cleave at specific DNA sequences, were used to insert the desired eukaryotic DNA sequence into a bacterial plasmid. Other enzymes, DNA ligases, were used to complete the recombinant bacterial plasmid that could be used to transform the host bacteria in order to propagate the eukaryotic gene in the bacteria. This allowed isolation of larger amounts of the eukaryotic DNA. Other techniques, including Sanger sequencing, were developed to determine the sequences of the various DNAs.[69] Initially readouts from Sanger sequencing had to be carried out on large electrophoretic gels with exposure to radiographic films. Later, sequencing devices were developed that separated the terminated DNA fragments by capillary electrophoresis and detected DNA fragments by fluorescence spectroscopy. This automated sequencing greatly increased the ability to carry out DNA sequencing in the 1990s.

Another technique that proved to be of great advantage to molecular biologists was polymerase chain reaction (PCR). Kary Mullis was a chemist working on short oligonucleotide sequences as probes for longer DNA sequences at the Cetus Corporation in Emeryville, California. One late night in 1983 when Mullis was on a date with a girlfriend who was also a chemist at Cetus, he had a brainstorm to use the bracketing short oligonucleotides to carry out multiple replications of the longer sequences of DNA. In developing this idea, Mullis used a DNA polymerase from the thermophilic bacteria Thermus aquaticus that had the unusual ability to survive temperatures above 75° Celsius. Mullis developed a technique for cycling the temperature to separate DNA strands after a replication process.[70] This allowed for multiple DNA replications for each cycle. Mullis's technique was enormously successful, allowing scientists to synthesize large amounts of DNA of a particular sequence. PCR machines were developed that could be programmed to automatically carry out the desired multiplication of DNA from a small amount of DNA. While Mullis became famous for this invention, and in 1993 was eventually awarded a share of the Nobel Prize in Chemistry, he commented that all the fame never made up for the breakup with his girlfriend.[71]

Much of molecular biology in the 1980s and 1990s involved cloning either genes of interest or cDNAs derived from the expressed mRNA to better understand structure and function of these genes. Most frequently, the protein of interest was isolated and sequenced either partially or fully. Having at least a partial protein sequence allowed the researcher to synthesize small oligonucleotides and to then probe the corresponding gene or mRNA (cDNA). With the techniques of the time, as described above, the researcher could obtain the entire gene sequence, the sequence of the specific mRNA, and the complete sequence of the expressed protein. Often, the revealing of the full protein sequence gave clues to the structure/function and regulation of the protein. In most cases, chemically sequencing the entire protein has tended to be a laborious process, so cloning of the cDNA has been a much easier route to probing the full-length protein sequence.

As an illustration of sequence determination by molecular cloning of the mRNA, I give an example of a project that I was a part of in the 1990s. The p70 S6 kinase is a protein kinase involved in the regulation of protein synthesis on ribosomes that is regulated by growth factors such as EGF, PDGF, insulin, and others.[72] The project was to first purify the p70 S6 kinase protein from rat liver and obtain a number of short peptide sequences of the protein. These peptide sequences were used to synthesize the corresponding oligonucleotides of the p70 S6 kinase mRNA, and these "oligos" were used to carry out PCR analysis of a cDNA (polyA mRNA) library from H4 hepatoma cells. The selected specific p70 S6 kinase cDNAs were then cloned and sequenced to reveal the complete primary structure of the 2278 base pair p70 S6 kinase cDNA. This in turn revealed a protein open reading frame encoding a 525 amino acid protein with calculated molecular mass of 59,186 Da. The determination of the p70 S6 kinase protein sequence allowed comparison with a variety of other related protein kinases, and it also predicted a possible C-terminal pseudosubstrate domain that was postulated to regulate its enzymatic activity by means of phosphorylation. This study is somewhat representative of the probing of individual genes to determine the gene sequence or the corresponding protein sequence. Cloning of the derived cDNAs can also make possible the insertion into mammalian expression vectors and artificial expression of these proteins in other cells that don't normally express them. Transfection of these cloned genes into mammalian cells not expressing the protein of interest can allow the study of the effect of the protein of interest on the metabolism in the transfected cell.[73]

While these types of studies of individual genes are still done, the availability of whole genome sequences has facilitated the process. Now, given peptides of the protein of interest, corresponding oligos can be synthesized and used to identify the position of the gene in the genome. Having identified the position of the gene in the genome, the researcher then has the sequence available from the human or other genomes that have been previously sequenced in their entirety.

One of the first scientists who advocated sequencing the entire genome was Renato Dulbecco, a noted virologist, who wrote the following in a 1986 perspective in *Science* magazine:

> *We are at a turning point in the study of tumor virology and cancer in general. If we wish to learn more about cancer, we must now concentrate on the cellular genome. We are back to where cancer research started, but the situation is drastically different because we have new knowledge and crucial tools, such as DNA cloning. We have two options: either to try to discover the genes important in malignancy by a piecemeal approach, or to sequence the whole genome of a selected animal species. The*

> *former approach seems less formidable, but it will still require a vast investment of research, especially if the important genes differ in cancers of different organs and if they encode regulatory proteins. . . . I think that it will be far more useful to begin by sequencing the cellular genome. The sequence will make it possible to prepare probes for all the genes and to classify them for their expression in various cell types at the level of individual cells by means of cytological hybridization. The classification of the genes will facilitate the identification of those involved in progression.*[74]

While Dulbecco pointed to human genome sequencing as a benefit to the understanding of human cancers, many other biologists were of the understanding that studies in a broad range of human diseases would benefit from the sequencing of the human genome. Prior to 1986, a number of prominent scientists had also promoted the idea of sequencing the entire human genome. These included Robert Sinsheimer of the University of California, Santa Cruz; Charles DeLisi and David Smith of the Office of Health and Environmental Research (OHER) at the U.S. Department of Energy; and James Watson of the Cold Spring Harbor Laboratory. Promotion of the project by DeLisi, who was the director of the OHER, resulted in the project being incorporated into the 1987 federal budget and funded initially at $16 million, with an expected cost of $3 billion over fifteen years. James Watson was appointed as the head of the NIH National Human Genome Research Institute (NHGRI) Project Initiative in 1990, and he was succeeded by Francis Collins in 1993.[75]

The human genome comprises a total of approximately 3.1 billion base pairs over twenty-three pairs of chromosomes. The goal of the project was to sequence all protein and RNA coding regions of the human genome. Another goal was to sequence other genomes, including those of the mouse, rat, and fruit fly. For the human genome sequencing, DNA was obtained from a number of individuals. Thus, the final sequence was a mosaic of all the individual donors. The genomes were mapped for known markers and broken down to approximately 20,000 manageable sections of 150,000 to 200,000 base pair fragments.[76] These fragments were cloned into bacterial artificial chromosomes (BACs). The approximately 20,000 clones derived from the complete human genome were distributed for sequencing to twenty different laboratories that were part of the International Human Genome Sequencing

Consortium.[77] Sequencing was carried out uniformly by automated "Next Gen DNA sequencing" that could essentially be run out around the clock. Sequences were subject to quality control by independent labs. By June 26, 2000, a rough draft sequence of the genome was finished that had an error rate of 1 base pair for every 1,000 base pairs and approximately 150,000 gaps that were not sequenced.[78] Independently of the NHGRI, Celera Genomics under the leadership of Craig Venter carried out their own sequencing of the human genome using a more efficient next-generation sequencing technique.[79] In October 2004, a finished version of the human genome was published that had an error rate of 1 base pair for every 10,000 base pairs and only 341 gaps that were not sequenced. It was estimated that in the final version there were approximately only 20,000 to 25,000 protein-coding genes.[80]

It was anticipated that the publishing of the human genome would have ethical, legal, and social implications. Making patients' DNA sequences public could easily affect their legal rights, their ability to obtain health insurance, and their susceptibility to being discriminated against on the basis of possible genetic defects. For this reason, the Ethical, Legal, and Social Implications (ELSI) program was established at the NHGRI to ensure that in the age of widespread genomic sequencing, citizens' rights would be protected. Thus, a layer of protection has been added to all human DNA sequencing such that sequences will be available, but personal information such as names, Social Security numbers, or other personal information will not be revealed to the public.

In the last twenty years since the publication of the draft version of the human genome, there have already been many advances in science and medicine due to the knowledge derived from the human genome.

As an extension of NHGRI, The Cancer Genome Atlas (TCGA) Program was established in 2006 to describe all the mutations in twenty thousand primary cancer samples and matching normal samples for thirty-three cancer types.[81] This project has promoted a greater understanding of tumor development and is leading to more personalized therapies for these cancers.

Also, as part of the NHGRI, the Encyclopedia of DNA Elements (ENCODE) Consortium has been formed as a group of twenty-seven institutions across the

United States. Genome-wide studies are conducted in a variety of areas, including

(1) 3-D chromatin structure

(2) chromatin accessibility

(3) chromatin interactions

(4) methylome

(5) chromatin modifications

(6) transcriptome

(7) RNA binding

These studies enhance the knowledge of the genomes of a variety of species in terms of structure and functional characteristics beyond the basic sequences of the genes making up the genome.

Another NIH program that will promote personalized medicine in a wide variety of areas is the "All of Us" research program. In this initiative, genomic and other patient-specific data will be collected on one million individuals in order to gain insights into the following:

(1) Who is at risk of certain diseases?

(2) Which treatments work best for people of different backgrounds?

(3) How can we connect people with the right clinical studies for their needs?

(4) How can we use technologies that will keep us in good health?

This data will be made available to physicians to help them make better personalized medicine decisions for their patients and to avoid a "one size fits all" medical care. The program will also attempt to provide the best medical care to patients of specific backgrounds as opposed to relying on a trial-and-error approach.

At the announcement of the completion of the draft human genome sequence on June 26, 2000, President Bill Clinton made the following statement about the significance of the achievement:

Today's announcement represents more than just an epoch-making triumph of science and reason. After all, when Galileo discovered he could use the tools of mathematics and mechanics to understand the motion of celestial bodies, he felt, in the words of one eminent

> *researcher, that he had learned the language in which God created the universe.*
>
> *Today we are learning the language in which God created life. We are gaining ever more awe for the complexity, the beauty, the wonder of God's most divine and sacred gift. With this profound new knowledge, humankind is on the verge of gaining immense new power to heal.*[82]

I think that Clinton was correct in citing three results of knowing our complete genomes:

> *(1) There will be increased knowledge and comprehension of the biological world.*
>
> *(2) We will learn the language of how God created life.*
>
> *(3) We will develop new medicines and technologies to treat diseases.*

The aim to create new ways of treating disease was an obvious goal of the Human Genome Project. It was hypothesized that by discovering new drug targets, the project would enable new targeted and personalized medicines to be developed. Twenty years after the publishing of the draft sequence, some progress has been made in certain areas of medicine. However, to date we have not seen the expected transformation of the way medicine is done. In the area of cancer treatment, for example, we see a few cancer treatments that use tumor genomic sequencing to inform a personalized therapy. For the most part, cancer therapy still starts with a one-size-fits-all standard of care therapy that does not use the genetic information. Given enough time to develop new therapies, we will, I think, see the greater use of personalized therapy based on knowledge of individual genomes. However, this will take more time, possibly another twenty or thirty years.

Likewise, I think the understanding of the language by which God created living organisms will also take a lot of time. It will take more than just learning which genes make which proteins. It will take a comparison of genomes of different species and the understanding of which genes are important for new traits of evolving species. Some of this work has been done, and we can see the way in which God may have used evolution to accomplish his creative work. My only disagreement with Clinton's statement is that I see the language that God used to create life to be reflected in but not directly revealed by the scientific knowledge of the genome. By studying the genome, we do not come to a direct understanding of how God created life. However, the patterns that we see in the genome may be a reflection of God's nature in creation.

The function of the genome is not only to direct the expression of certain proteins in different tissues. The genome directs the entire process by which organisms develop from a single fertilized zygote to a fully formed adult. An early-stage embryo, as its cells multiply, goes through distinct stages to form a blastula, gastrula, and neurula. At present, most developmental processes from an embryo to the various stages of complex development are poorly understood in terms of what genes specify the various folding and differentiation processes. Further study of the genome will no doubt shed light on the principles by which the genome governs these processes of development.

∞

As we look at the overall patterns of gene expression in development and with comparison to other species, we will have a better idea of how those species arose. We may even see a reflection of God's creative activity in bringing about these species. It is important to understand that what we see in science is only a reflection of God's activity. I would not venture to say that we could directly see God's activity in creation. I would not even venture to say, as those who argue for intelligent design do, that the complexity of the structures argues for the existence of God. However, by looking at the majesty of structures and functions of the genomes in these organisms, we can see just a shadow of how the transcendent God might govern these activities through scientific principles.

Chapter 6:

Enzymology, Cellular Signaling, and Theology

Enzymology is the study of proteins that catalyze chemical reactions from reactants to products. A big step in the history of enzymology was establishing that enzymatic reactions could be carried out in cell-free lysates. Starting in 1893, Eduard Buchner isolated cell-free extracts of yeasts and was successful in generating alcohols from glucose in these extracts. He hypothesized that enzymes in the extracts were responsible for converting glucose to alcohol. This was the first step in showing that all the chemical reactions of the cell could be mediated by enzymes and would not require invoking of elements or forces outside the cell.[83]

The first enzyme to be isolated in pure form was urease, the enzyme that breaks down urea into ammonia and carbon dioxide. In 1926, James Sumner extracted urease from the jack bean *Canavalia ensiformis*.[84]

Sumner found that he could precipitate the enzyme with dilute acetone, and then filter the remaining dissolved proteins to yield the purified urease crystals on the filter. While some scientists challenged the claim, another scientist, John Northrop, crystallized pepsinogen (the precursor of pepsin), trypsin, chymotrypsin, and carboxypeptidase by similar methods, thus confirming Sumner's findings.[85] All of the areas of enzymology have relied on protein purification to isolate enzymes and relied on the establishing of in vitro enzyme assays to quantify enzyme activity.

In the last hundred years or so, the study of enzymes has branched off into a number of areas, including the following:

(1) cellular metabolism

(2) enzyme mechanisms[86]

(3) enzyme kinetics[87]

(4) enzyme structure[88] *(see chapter 4 for X-ray crystallographic studies)*

Of these areas, I will emphasize mostly the area of mammalian cellular metabolism. The reader can consult the references cited above to explore these other areas of enzymology.

Studies over the last fifty years have contributed greatly to the understanding of metabolic interrelationships between various mammalian tissues. These studies are closely related to subsequent work in cellular signaling, the process by which extracellular hormones signal to activate specific types of intracellular metabolism. While I do not attempt to comprehensively cover all areas of intermediary metabolism, a number of representative areas will be touched on.

The study of cellular metabolism in mammals has been important because essentially all physiological functions in the various tissues are undergirded by metabolic pathways modulated by specific enzymes. For instance, the liver is a tissue that processes nutrients coming from the intestine and is able to make metabolites available to other tissues as needed. Nutrients consist primarily of lipids, carbohydrates, and proteins. If we always ate exactly the components that we needed, we would not need such an elaborate system. However, the liver through its array of enzymatic pathways is able to interconvert various nutritional components into what is needed by the other tissues. This is why, for instance, when you go on a low-carbohydrate diet, you don't immediately die for lack of energy. Even though carbohydrates are usually the major source of energy for mammals, in the absence of carbohydrates, the liver can use lipids and protein as a source of energy. On the other hand, in the absence of lipids or proteins, available carbohydrates can be converted to these components. The only exception would be for certain essential amino acids or lipids, which cannot be produced by most mammals. These essential components are required to be provided in the diet. For instance, for humans, essential amino acids include leucine, isoleucine, lysine, methionine, phenylalanine, threonine, tryptophan, valine, histidine, and arginine.[89] The understanding of these pathways has given a basis for determining essential dietary requirements. However, these metabolic pathways make mammalian system versatile under various dietary conditions to finely regulate the components needed for the organism to function.

Approximately two-thirds of dietary carbohydrates are removed by the liver and stored as glycogen. As needed, glucose is made available to other tissues including muscle, fat tissue, and the brain. Once metabolized through glycolysis and the citric acid cycle, glucose is a source of energy for muscles and the brain. Glucose can also be transported to fat tissue, where it is stored as glycogen or converted to triglycerides for storage. The liver also functions to make free fatty acids available either directly in the blood or as complexed with very low-density lipids (VLDLs). Fatty acids can also be metabolized by muscle tissue through Acetyl CoA feeding into the citric acid cycle. The interdependence of these pathways is critical to the functioning of the organism and is mediated by hormone action.

As an illustration of hormonal regulation that keeps blood components under control, consider the regulation of blood glucose levels by the hormones insulin and glucagon. Blood glucose is required to be maintained at a fasting level of 80± 20 mg/dL in order to maintain brain function and to supply energy to other tissues, such as skeletal muscle and the heart. After a meal, an increase in dietary carbohydrates and blood glucose level (hyperglycemia) stimulates an increase in insulin secretion by the β-cells of the pancreatic Islets of Langerhans. Insulin affects a multitude of tissues. Liver cells are stimulated to increase active intracellular transport of glucose and increase glycogenesis (storage as glycogen in those cells). Likewise, glycolysis is increased in these cells, and gluconeogenesis is decreased in these cells. In other tissues, such as muscle, brain, and adipose cells, active transport of glucose is also increased. In most of these tissues, glycolysis and glycogenesis are increased. In some cases, protein synthesis is also increased as energy sources increase. In fat cells, lipogenesis increases, and lipolysis decreases. Action of insulin through the insulin receptor will be discussed in the section on cell signaling. Increase of blood glucose from a meal also causes a decrease in secretion of glucagon by the α-cells of the pancreas. Effects of decreased glucagon are primarily to decrease gluconeogenesis and glycogenolysis in the liver. These actions of decreased glucagon aid in the bringing back of blood glucose from the higher level to the homeostatic level. Conditions of hypoglycemia (low blood sugar) cause decreased insulin secretion by β-cells of the pancreas and increased secretion of glucagon by the α-cells of the pancreas. Effects on tissues are essentially the opposite of those shown above for hyperglycemia. Overall, the effect is to increase the blood glucose level to bring it back to the homeostatic level.

The study of cellular signaling in the mammalian context is about how a hormone or growth factor in the blood conveys a signal to modify the intracellular metabolism via a transmembrane receptor. The targets of these signals may be metabolic enzymes or proteins that stimulate cellular processes such as gene transcription, protein synthesis, or cell survival. These added layers of regulation are essential to the coordination of metabolic processes in higher organisms.

The field of cellular signaling grew out of the work of biochemists and enzymologists in the latter half of the twentieth century. One particularly influential scientist at the forefront of this field was Edwin Krebs.[90] Krebs received an undergraduate degree in chemistry but decided to pursue a medical degree at Washington University in Seattle. Krebs finished his MD degree while serving as an officer in the Navy during World War II. At Washington University, Krebs had significant exposure to basic biochemistry research. He developed a strong association with Carl and Gerty Cori in the Department of Pharmacology. At the end of the war, Krebs had to wait two years before entering a medical residency, so he decided to spend the time as a postdoctoral fellow in the Cori lab from 1946 to 1948. This turned out to be a fateful decision in Krebs' future scientific career.

Krebs' first project in the Cori lab would be in working with the enzyme glyceraldehyde 3-phosphate dehydrogenase (GPD), which converts D-glyceraldehyde 3-phosphate to 1,3-bisphospho-D-glycerate. Krebs observed conversion of bound NADH to NAD^+, what he referred to as NADH-X. He was unable to unravel the full significance of this reaction, but other workers later showed that this NADH-X was an intermediary form of bound D-glyceraldehyde 3-phosphate.[91] A subsequent reaction released the product, 1,3-bisphosphoglycerate from glyceraldehyde 3-phosphate dehydrogenase.

In 1953, a new member of the Cori lab, Edmond Fischer, became interested in another enzyme, glycogen phosphorylase, which was involved in the breakdown of glycogen to glycogen plus glucose 6-phosphate. At this point, Krebs and Fischer entered into a collaboration on phosphorylase that would last for fifty years.[92] It was known that there were two forms of phosphorylase, a (the active enzyme) and b (the inactive enzyme). Furthermore, conversion from the b to the a form was supposed to involve a prosthetic group-removing enzyme (known today as a phosphatase). Krebs was also interested to explore the conversion of the b form back to the a form, a process now known to require ATP and a protein kinase. As a starting point, Krebs attempted to isolate the a form of phosphorylase using a previous method established in the Cori lab for rabbit muscle phosphorylase a. All of the phosphorylase purified turned out to be in the inactive, b form. On examining the protocols, Krebs found that the only difference in the protocols was substituting a centrifugation step for a filtration step for the crude homogenate. When Krebs switched back to the filtration step, all purified phosphorylase was now in the active, a form. Thus, it appeared that some process involving the filtration of the extract led to a phosphorylation converting phosphorylase b to phosphorylase a. It turned out that the filter provided Ca^{2+} ions that activated another kinase, later termed phosphorylase kinase. In the presence of ATP, phosphorylase kinase (activated by Ca^{2+}) converted the inactive phosphorylase b to active phosphorylase a. In the reverse direction a phosphatase converts phosphorylase a to phosphorylase b (see figure 6.1).

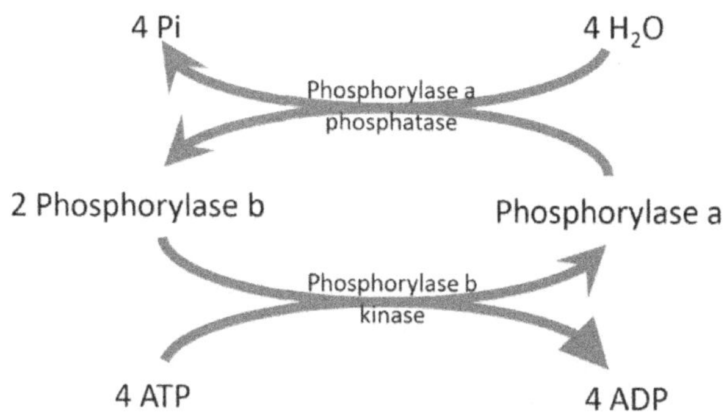

Figure 6.1. Conversion of phosphorylase b to phosphorylase a (adapted from J. M. Orten and O. W. Neuhaus, Human Biochemistry [St. Louis: C.V. Mosby, 1982], 254).

This was also confirmed by the work of Earl Sutherland, another member of the Cori lab, who found the same conversion of phosphorylase a to b in the liver. Sutherland later found that in the liver, the hormone epinephrine stimulated intracellular production of cyclic AMP. This first example of an intracellular hormone mediator was found to activate a protein kinase that they termed protein kinase A (PKA). Figure 6.2 shows the signaling cascade from extracellular epinephrine in the blood to breakdown of glycogen in the liver and skeletal muscle. This was the first example of a protein kinase/phosphatase signaling cascade. As it turned out, the vast majority of intracellular signaling cascades in mammals are controlled by a similar phosphorylation and dephosphorylation of serine, threonine, or tyrosine amino acid residues of proteins. Thus, while these other signaling cascades were all modulated by different enzymes, they all shared activation and deactivation mechanisms by reversible ATP-dependent protein phosphorylation. Because Krebs and Fischer were on the forefront of this work on protein phosphorylation/dephosphorylation, they were jointly awarded the Nobel Prize for Physiology or Medicine in

1992.[93]

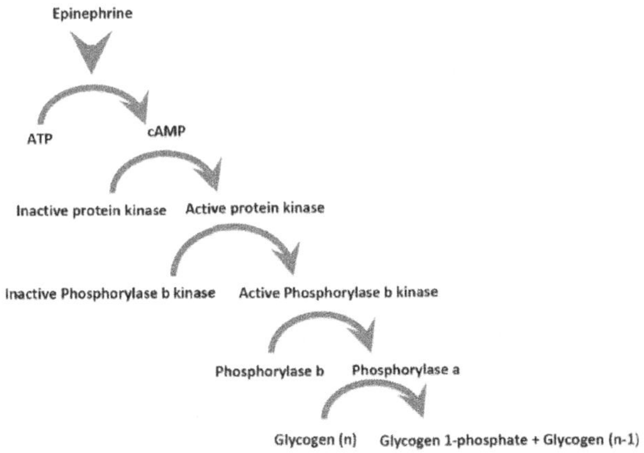

Figure 6.2. Epinephrine stimulates a glycogen phosphorylase cascade (adapted from Orten and Neuhaus, Human Biochemistry, 255).

Since the 1970s, there have been a number of different contributions to fill in gaps in these signaling pathways and to expand on their downstream functions.

In the mid-1980s, Robert Lefkowitz of Duke University discovered and cloned the gene for the receptor for which epinephrine signals to β-adrenergic functions (i.e., glycogenolysis, lipolysis, and others).[94] This receptor, the β-adrenergic receptor, was a member of the G-protein Coupled Receptor (GPCR) family.

These receptors are characterized by a serpentine polypeptide chain with seven regions passing through the plasma membrane. Binding of a ligand such as epinephrine to the GPCR on the outside of the cell stimulates a release of GDP from the associated G-Protein α-subunit on the inside of the cell and the binding of GTP in its place, leading to α dissociation from the β and γ subunits. It has been shown that the free GTP-bound α subunit could bind to another effector molecule, which is adenyl cyclase in the case of the β-adrenergic receptor. This stimulation of adenyl cyclase results in production of intracellular cyclic AMP (cAMP), leading to activation of Protein Kinase A (PKA) (see figure 6.3). Active Protein Kinase A then leads to phosphorylation of targets and to such processes as glycogenolysis (figure 6.2) in the liver and lipolysis in fat tissues.

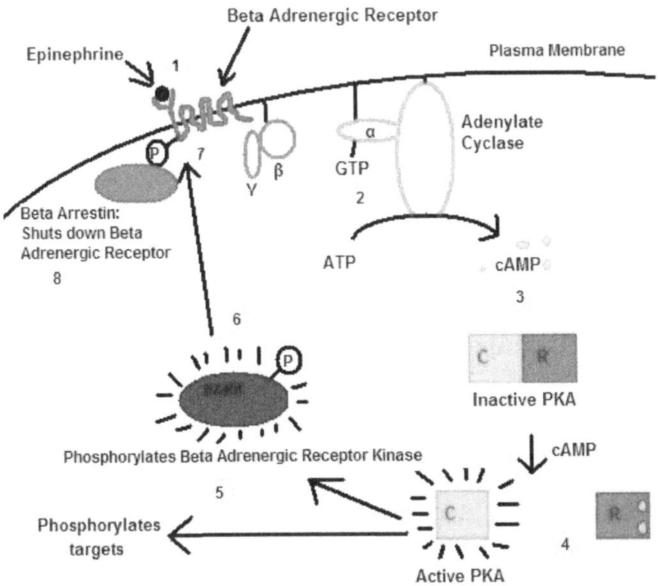

Figure 6.3. Stimulation of beta adrenergic receptor by epinephrine (see Kitra101, "File:Beta Adrenergic Receptor Kinase.JPG," Wikimedia Commons, last edited [08-19-2020], [https://commons.wikimedia.org/wiki/File:Beta_adrenergic_receptor_kinase.JPG]).

Hydrolysis of α-subunit-bound GTP to GDP results in reassociation of α, β, and γ subunits of the G-protein and a return to the deactivated state of the receptor. Subsequent work has shown that the GPCR family of receptors includes about one thousand members, signaling a multitude of hormones and neurotransmitters. It is estimated that 30 to 50 percent of today's prescribed drugs are targeted to GPCRs. Medications include antihistamines, antiulcer medications, and cardiovascular medications.[95]

For the hormone insulin and other growth factors (EGF, PDGF, c-kit, and others) a different type of receptor signaling has been seen. In the late 1970s and early 1980s, Raymond Erikson and Tony Hunter showed that in addition to serine and threonine residues, tyrosine residues on proteins could also be phosphorylated.[96] Hunter and Cooper then showed, in A431 tumor cells, that the growth factor, EGF, bound to a dimeric transmembrane protein termed the EGF-receptor, resulting in autophosphorylation of tyrosine residues on this receptor.[97] This led to the findings that a number of other growth factors (PDGF, c-Kit) and insulin also signaled through related dimeric transmembrane tyrosine kinase receptors that were activated via tyrosine autophosphorylation. Downstream signaling via the receptor tyrosine kinases proved to be relayed by serine and threonine kinases. A variety of these kinases, including MAP kinases, PI3-kinases, and mTOR kinases, have been shown to be implicated in downstream signaling. These classes of kinases have been shown to stimulate functions including transcriptional activation, cell survival, and protein synthesis, respectively, in a wide range of cells.[98] Figure 6.4 shows how extracellular EGF stimulation of the EGF receptor tyrosine kinase leads to stimulation of nuclear transcription that is mediated by cytoplasmic MAP kinases and other related Ser/Thr kinases.

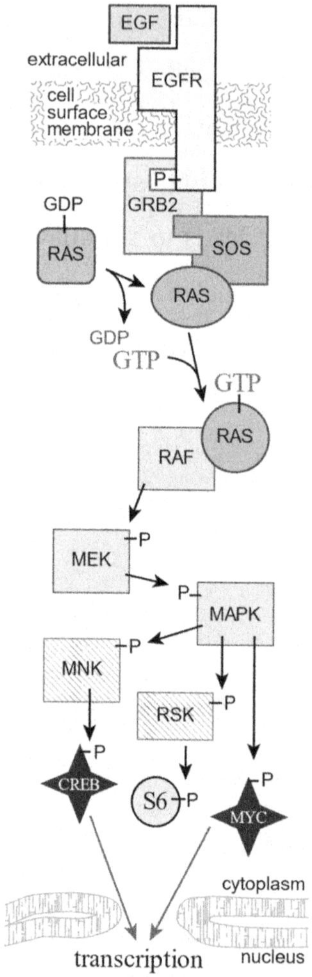

Figure 6.4. Extracellular EGF stimulation of EGF receptor tyrosine kinase leading to nuclear transcription mediated by MAPK and other Ser/Thr kinases (see Fred the Oyster, "File:MAPKpathway diagram.svg," Wikimedia Commons, [08-19-2014], [https://commons.wikimedia.org/wiki/File:MAPKpathway_diagram.svg]).

The above discussions are meant to give a broad overview of hormone-mediated metabolism in higher organisms. The point is to show the ground-floor metabolic pathways and functions that are regulated by the upper-floor hormones and growth factors via intracellular signaling pathways in the various tissues of higher organisms. It is also helpful to see in a general way how these two floors are connected through various regulatory modes of signaling, including receptor binding and protein phosphorylation.

When we step back to get an overview of the vast diversity of mammalian metabolism, and the regulation of that metabolism, we see a coordination of perhaps thousands of enzymes in concert with a multitude of hormones and growth factors. Each of the enzymes is tuned with specific binding constants to modulate the conversion of various reactants to products. With such an array of enzymes and regulators, one cannot fail to conjecture whether these systems came about by chance, or by some grand design of a creator. At least three different scenarios are possible:

> *(1) All enzyme systems were formed individually de novo for each species under the direction of a divine creator.*
>
> *(2) Enzyme systems evolved by chance as species evolved from common precursors, according to Darwinian principles in the absence of a divine creator.*
>
> *(3) Enzyme systems evolved as species evolved, according to principles of natural selection but under the governance of a divine creator.*

When we look at the average differences in sequences of proteins in various species, we see a relatedness that parallels the proposed evolution of the various organisms from common precursors. For instance, in the family tree of the cytochrome Cs, we see average sequence differences in twenty-one species, suggesting that protein sequence relatedness parallels species relatedness.[99] Such a correlation would argue against all of the cytochrome Cs arising independently de novo as in scenario 1. Arguments for common descent of species along with common evolution of gene products have been made based on comparisons of sequences within genomes. Francis Collins, in his book *The Language of God*, gives a number of examples of

DNA sequence similarities among related species that argue strongly for common descent of these species. For instance, orders of genes are often similar on homologous chromosomes when the human and the mouse genomes are compared. Patterns of nonfunctional DNA—so-called ancient repetitive elements (AREs)—are also found in similar positions when the mouse and the human genomes are compared. As Collins clearly states: "Unless one is willing to take the position that God has placed these decapitated AREs in these precise positions to confuse and mislead us, the conclusion of a common ancestor for humans and mice is virtually inescapable. This kind of recent genome data thus presents an overwhelming challenge to those who hold to the idea that all species were created ex nihilo."[100]

Thus, there is a fair amount of certainty that enzymes evolved as species evolved from common precursors, ruling out scenario 1. However, distinguishing between scenarios 2 and 3 is more difficult. The meaning of such terms as "chance events" and "random events" has been widely debated. Some have posited that the observance of events in variation and mutation of genes that appear randomly are evidence for the absence of God in determining the outcome. Assortments of genes occur through Mendelian genetics and genetic recombination events during meiosis by apparently random processes.

Deborah and Loren Haarsma, in their book *Origins*, have stated that "when scientists say that something is *random*, they mean that the outcome is unpredictable." They give the example of calculating the outcome of the roll of a pair of dice. The outcome can never be determined exactly but can be predicted statistically.[101] From the point of view of one who believes in a transcendent God, this is not to say that God does not govern the outcome. However, if God is involved in the process of evolution, would we not expect to see evidence of nonrandom processes? If God is truly transcendent and yet the one who governs the laws of natural selection, we would expect to see genetic behavior following seemingly random patterns. This is not an argument for his absence. This only means that he has ordered certain laws under which these genetic phenomena follow statistical and regular patterns. It also does not prevent him from very occasionally deviating from those regular patterns—what we would term an intervention.

Likewise, we see evidence for design in the metabolic regulatory pathways. Almost all the regulatory pathways between the two floors of metabolism and transmembrane hormone receptors utilize protein phosphorylation as a means of regulation. Transmembrane receptors are of predominantly two types, GPCRs (figure 6.3, above) and receptor tyrosine kinases (figure 6.4, above). What are the chances that a random process of pathway development would focus primarily on these limited mechanisms of regulation? It should be noted that the evidences for design here are not meant to be examples of intelligent design. The concept of intelligent design was developed by Phillip Johnson in the 1980s to promote a supernatural designer that substituted for Darwinian evolution, which utilized a gradual, seemingly random, step-by-step, trial-and-error building of complex systems.[102] As will be seen in the next chapter, on evolution, there is another option, termed theistic evolution, that employs the seemingly random mechanism of Darwinian evolution, but under the governance of a transcendent god.

∞

The evidences for design in metabolic pathways noted here do not supersede the principles of evolution that show a step-by-step variation and selection by survival of the fittest. The patterns noted here are a mere reflection of a supernatural designer who governs the principles of the evolutionary process. They do not prove the participation of that designer or substitute for Darwinian principles operating in evolution. As will be noted in the chapter on biological evolution, it is critical to formulate a concept of theistic evolution that does not limit the principles of Darwinian evolution, but also does not limit the power of a transcendent God in governing those principles. The basic idea is that God uses methods that could be viewed as random choices, but in the end, the patterns that are seen in the structures and pathways are in some ways not random. This reflection of the divine designer is what we see in nature if we are careful to view the broad overview of creation in biological systems. It does not totally refute the naturalistic concept of atheistic evolution, but it leaves open the possibility of a divine creator who governs the principles of evolution to create the patterns in nature that we see. In mammalian metabolism we see a majestic tapestry of interlocking pathways and regulators that maintains the carefully tuned functions of the organism. Considering such a massive tapestry would cause us to consider the participation of a transcendent god as designer and creator.

Chapter 7:

Evolutionary Biology and Theistic Evolution

> *There is grandeur in this view of life, with its several powers, having been originally breathed by the Creator into a few forms or into one; and that, whilst this planet has gone cycling on according to the fixed law of gravity, from so simple a beginning endless forms most beautiful and most wonderful have been, and are being evolved.*
>
> Charles Darwin, The Origin of the Species (1859)

Many a scientist has started on the path to rational knowledge with the understanding that there is a Creator God and has finished that pursuit with the conclusion that there is not such a Creator God. Such was the case with Charles Darwin. In finishing his book *The Origin of the Species*, he credited a Creator God as having a role in the evolution that he observed. However, at the end of his life, he became an agnostic, finding it preferable to abandon his faith in God rather than to abandon his faith in his scientific findings. Many, like Darwin, have found that faith in a transcendent God and faith in science are not compatible. George Gaylord Simpson, the noted biologist, said about religion and science: "Each has its own sphere, and friction arises only when either one is invaded by the other."[103] An alternate view is given by Theodosius Dobzhansky, a prominent evolutionary biologist: "Does evolutionary doctrine clash with religious faith? It does not. It is a blunder to mistake the Holy Scriptures for elementary textbooks of astronomy, geology, biology, and anthropology. Only if symbols are construed to mean what they are not intended to mean can there arise imaginary, insoluble conflicts."[104] So, to address Simpson's objection, we need not keep these two spheres separate, but need to be very conscious of the intent of the writers and the interpretation of their language. This will allow us as scientists who believe in a transcendent God to have a harmonious understanding of how these spheres interact. It is on this basis that the concept of theistic evolution becomes a firm foundation to build one's worldview. There need be no fear on the part of the scientist that the spiritual and the physical spheres will conflict with one another.

Charles Darwin was born into a middle-class English family on February 12, 1809.[105] As a young student at Edinburgh University, Darwin was not on track to change the world of biology. He initially attempted a major in medicine, but found anatomy boring and was disgusted by the practice of dissection. After he transferred to the University of Cambridge in 1828, Darwin's father suggested that he consider a career in the clergy. In the end, Darwin felt that he was not orthodox enough in his Christian belief to become a priest in the Church of England. Although Darwin admitted to wasting a lot of his time in lectures at Cambridge, he was also developing interests in natural history, geology, and entomology, collecting beetles and hunting birds. One particularly influential professor, John Stevens Henslow, who was well versed in botany, entomology, chemistry, and geology, mentored Darwin personally and made these subjects come alive to him. Darwin was as impressed by Henslow's moral qualities as he was by Henslow's approach to science. Henslow was also instrumental in getting Darwin a position as a naturalist on the HMS *Beagle* as it embarked on a five-year voyage around South America. This journey, from December 1831 to October 1836, would change the course of Darwin's career as a naturalist. His main task as naturalist was to document wildlife and geological history throughout the journey. While thriving as a scientist, he was seasick most of the time and survived on dried raisins and hard tack. He had no intentions to develop a theory of the origin of living species; he considered it his role to document and classify the wildlife of South America in their habitats. He also would examine fossils to compare preexisting species with the present wildlife. It was as if he were back at Cambridge collecting beetles, but his world had been expanded to the coasts of South America, with its vast diversity of species. Most of the samples that Darwin collected and classified were packed up and sent back to Great Britain to be analyzed by other naturalists.

Darwin returned to England in 1836. As he and his naturalist coworkers studied and classified his samples, they drew two major conclusions. The first concerned fossils of giant mammals in Argentina. One fossilized head (whose owner was later termed a Toxodon) seemed to indicate an animal the size of a hippopotamus, but from the teeth and other anatomy the animal was determined to be related to the rodent order. Darwin also brought back from Argentina large plate-like armor derived from a giant armadillo creature termed Glyptodon, known to exist during the Pleistocene epoch. Darwin and his colleagues concluded that these giant mammals existed in ancient times, before the smaller mammals of current times.

Second, the scientists drew conclusions about birds Darwin had brought back from the Galapagos Islands. According to Siddhartha Mukherjee, Darwin had classified these as variations of wrens, warblers, blackbirds, and "Grossbeaks."[106] The ornithologist John Gould looked at these diverse birds and realized that they were all finches. Each inhabited a particular island of the Galapagos and had adaptations specific to that island.

These two findings drove Darwin to try to make sense of how these current varieties of finches and mammals might have arisen from past history. It occurred to him that all of the finches might have arisen from one ancestral mating pair of finches that years ago had migrated to one of the islands. That pair then reproduced, and its descendants spread to the other islands. As time passed, he hypothesized, the finch population of each island adapted to the environment of that island via useful traits that developed over the course of generations. Likewise, the giant mammals produced smaller variants that resulted in the current small-sized species. These smaller variants survived because they were better suited to their environments than were the larger mammals. While Darwin knew nothing of how a species would pass on its hereditary information to the next generation, he hypothesized that such a variation that occurred in one parent could be passed on to subsequent offspring and preserved for subsequent generations. The question then became "Why would certain variations be favored over other variations?" For an answer, Darwin turned to *An Essay on the Principle of Population*, by Thomas Malthus.[107] Malthus hypothesized that a population would continue to grow until it exceeded the limit of food and water resources; then it would decrease by death or lower reproduction. Darwin concluded that in a mixed population of varied species, the variant best adapted to the environment would be the one to survive, because it would be able to make the best use of the available resources. He coined the term "survival of the fittest" to describe the force that would lead to the development of new species adapted to a new environment. Darwin spent more than fifteen years developing his theory about how different species would originate from a common precursor, forming a tree-like genealogy.

In 1855, Darwin received a letter from Alfred Russel Wallace, who had developed a similar theory of species origination. Wallace had observed species in Malaysian islands that had been gradually separated by channels of water creating smaller islands. Wallace had also read Malthus' article and had concluded, like Darwin, that species would originate by variations followed by survival of selected species based on the best fitness to the environment. Seeing that both had stumbled on essentially the same theory, Darwin sought advice from his geologist friend Charles Lyell. Lyell suggested that both Darwin and Wallace present papers on the subject at the Linnean Society in London. In 1858, both papers were presented but not with much notice by the members of the Linnean Society. Apparently, these hard-core naturalists were already stuck on their own theories of species origination. In 1859, Darwin published a fuller explanation of his theory entitled *The Origin of the Species by Means of Natural Selection.* In contrast with the 1858 paper, the book generated great interest, with all twelve hundred copies of the first printing selling on the first day.

The debate over evolution and creation has been, without a doubt, one of the most controversial topics in circles of spiritual and scientific thought. A variety of flavors of biological origins have been proposed:

(1) Young Earth Creationists (YECs): Based on a calculation by Anglican bishop James Ussher (1581–1656) based on literal interpretation of the book of Genesis, YECs believe that God created the universe in 4004 BC (approximately 6000 years ago) in six twenty-four-hour days, creating Adam and Eve on the sixth day.

(2) Old Earth Creationists (OECs): Based on geological and astronomical evidence, OECs believe that the universe is relatively old (i.e., about fourteen billion years). However, they also believe that life forms were created by God de novo over the approximately four billion years of the earth's history.

(3) Intelligent Design (ID) Advocates: While there are several versions of ID, most ID advocates hypothesize that creation may have been in gradual steps, but that the complexity of living forms cannot be explained by Darwinian evolution (i.e., random mutations leading to higher structures). ID advocates tend toward believing that complex structures were created by God de novo through some creative event.

(4) Theistic Evolutionists (TEs) or Evolutionary Creationists (ECs): TEs or ECs believe the universe is about fourteen billion years old, as indicated by astronomical evidence, and they believe life has evolved on earth for about four billion years, predominantly through Darwinian evolution. However, they do not rule out the intervention of a divine being in facilitating the creation of life forms through what might appear to be a random selective process driven by survival of the fittest.

(5) Classical Atheistic Darwinian Evolutionists (CADEs): Believers in naturalistic or atheistic evolution favor an age of the universe based on astronomical evidence (i.e., about fourteen billion years) and believe the earth is about four billion years old. They

> further suppose the evolution of living forms over the age of the earth to be only due to random mutation and natural selection based on survival of the fittest, not due to any divine intervention.

Recall that genetic and biochemical sequence information greatly favors the theory that living species evolved from common descendants as proposed by Darwin. So, we can set aside the young earth creationists (YECs) and the old earth creationists (OECs) for another day. The question then becomes "Can we reconcile the existence of a transcendent God with descent of living beings from common precursors?" As has been pointed out by Ted Davis in his recent BioLogos post on theistic evolution,[108] there are a range of proponents of this middle ground, and it is important to appreciate the distinctions.

To help readers understand these distinctions, figure 7.1 indicates the range of the evolution-creation spectrum. To the far left we see adherents of purely random atheistic evolution (classical atheistic Darwinian evolutionists [CADEs]).

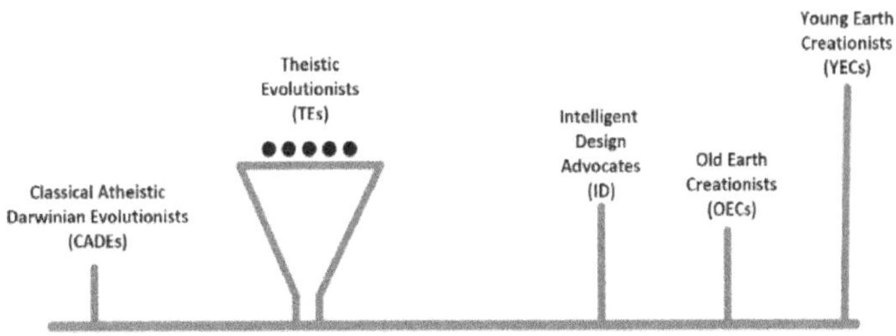

Figure 7.1. Theistic evolution within the range of the creation-evolution spectrum.

To the far right we see the holders of purely creationist views (young earth creationists [YECs] and old earth creationists [OECs]). Intelligent design (ID) advocates tend toward the right, as, for the most part, they deny the validity of Darwinian evolution in explaining the origin of complex structures. This leaves the middle territory of theistic evolution, which covers a somewhat diverse, but contiguous range of positions. The five variations within TE—each, in its own way, affirming the intervention of God in the evolutionary process—are BioLogos (proposed by Francis Collins), classical TE (William Whewell), modernist TE (Rudolf Bultmann and others), orthodox TE (John Polkinghorne and others), and punctuated interventional TE. These five approaches may be described further as follows:

(1) BioLogos: In *The Language of God*, Francis Collins described his version of theistic evolution, which he termed BioLogos. He starts with the presumption that "the universe came into being out of nothingness approximately 14 billion years ago."[109] He then states that the universe was precisely tuned for the conditions of life—presumably, by God, though he does not mention God. While the mechanism of the origin of life is not known, the process of evolution is thought to be sufficient to bring about the diversity of life forms. Furthermore, he states that "once evolution got underway, no special supernatural intervention was required."[108] While he affirms the origin of humans as through evolution from the great apes, he points out that, contrary to the atheistic evolutionary explanation, humans have a unique spiritual nature, including the awareness of "the Moral Law" and the need to search for God in a religious sense.[108]

(2) Classical TE: Like Collins, English philosopher and historian William Whewell (1794–1866) emphasized the creation of the world through establishment of natural law (i.e., fine-tuning of the universe) as opposed to any particular divine intervention: "But with regard to the material world, we can at least go as far as this—we can perceive that the events are brought about, not by insulated interpositions of divine power, exerted in each particular case, but by the establishment of general laws."[110]

(3) Modernist TE: Liberal theologians have tended to emphasize the immanence of God in creation to the exclusion of his transcendence. Likewise, they have tended to ignore the reality of such historical events as the virgin birth of Jesus and his bodily resurrection. Consider as an example, Rev. Harry Emerson Fosdick, who was a pastor at New York's First Presbyterian Church in 1922 and delivered a sermon entitled "Shall the Fundamentalists Win?" In it, he questioned a number of orthodox Christian beliefs including the virgin birth of Jesus, as indicated here:

Here, for example, is one point of view that the virgin birth is to be accepted as historical fact; it actually happened; there was no other way for a personality like the Master to come into this

> *world except by a special biological miracle. That is one point of view, and many are the gracious and beautiful souls who hold it. But side by side with them in the evangelical churches is a group of equally loyal and reverent people who would say that the virgin birth is not to be accepted as an historic fact.*[111]

For such theologians, in rationalizing scripture and the emerging science, they ended up calling into question statements of faith, such as the Nicene Creed, that have been the standard of Christian belief since the founding of the church. The real problem with Fosdick's questioning of the virgin birth is the clear statement in the gospels of Matthew and Luke that the birth of Jesus was the result of a miraculous conception (Matthew 1:18–25; Luke 1:26–35). Matthew in particular states: "When His mother Mary had been betrothed to Joseph, before they came together she was found to be with child by the Holy Spirit. And her husband Joseph, since he was a righteous man and did not want to disgrace her, planned to send her away secretly" (Matthew 1:18–19). If we are to take the gospels of Matthew and Luke as historical accounts, as most scholars do, we have to conclude that somehow these writers got their facts wrong. Numerous other theological incongruities also ensue concerning Jesus' divine nature, accounts recorded later in the Gospels, and the inspiration of scripture. However, the point to be made is that these miracles recorded in scripture—that we as Christians take as a matter of faith—are outside the realm of science. In harmonizing science with faith, we do not align the miraculous side by side with what we come to know by the scientific method. In fact, if God is truly transcendent, we can expect that miraculous events in the Bible will be contrary to what we observe in the natural world by the scientific method. Both are evidential realities, but according to their own standards. As Luke records, "Mary said to the angel, 'How will this be, since I am a virgin?'" (Luke 1:34). It appears that even Mary understood that the birth of Jesus was not a natural event. Harmonizing the supernatural and science means accepting both as they were meant to be interpreted. We cannot fit them both into a naturalistic mindset. Were we to do that, we would be minimizing the transcendence of God. Thus, the modernist TE approach makes an attempt to harmonize science and faith, but at the expense of a fuller appreciation of the divine nature.

(4) Orthodox TE: The scientist-theologian John Polkinghorne ties the creation of the universe to the Christocentric concept of Jesus suffering for our sins and dying on Calvary: "The insight of the Crucified God lies at the very heart of my own Christian belief, indeed of the possibility of such belief in the face of the way the world is. But this can only really be so if God is indeed truly present in that twisted figure on the tree of Calvary. Only an ontological Christology is adequate to the defense of God in the face of human suffering. God must really be there in that darkness."[112] Similarly, botanist Asa Gray (1810–88) states: "Revelation culminated . . . in the advent of a

Divine Person, who, being made man, manifested the Divine Nature in union with the human," and "this manifestation constitutes Christianity."[113] It is presumed by this underpinning of the Christological statements of faith that Christ was in fact creator of the universe and that it was this same Christ who suffered and died for our sins.

(5) Punctuated Interventional TE: This version of theistic evolution attributes the vast majority of diversity of living forms to Darwinian evolution through survival of the fittest for a particular environment. It also posits, however, that rare instances that would otherwise be interpreted as random occurrences are biased by divine intervention. As in other versions of TE, in this variation God uses the

Darwinian evolutionary principles to bring about life forms. However, because evolution is a sum total of individual reproductive events, these events cannot be subjected to investigation under controlled conditions. This principle of rare and noninvestigable intervention stands in contrast to BioLogos, which states that "once evolution got underway, no special supernatural intervention was required." Contrary to BioLogos, it is not the requirement of supernatural intervention that is questioned here. Moreover, in punctuated interventional TE it is asserted that God in his sovereignty reserves the freedom to intervene in conjunction with the overall unfolding of natural laws of evolution that he has established. These interventions, while exceedingly rare, can influence the course of evolution in a nonrandom way and may have a big influence over the types of life forms that predominate, even though the overall process is driven by Darwinian evolution.

The common threads to all of these approaches to theistic evolution are, first, that Darwinian evolution is the operative principle in the origin of variation in living forms, but God remains active to some degree in this process, and second, that the involvement of God in this process is not open to scientific investigation. Because God is postulated to be transcendent and dwelling in a domain that is beyond the three-dimensional realm that scientists have access to in carrying out experiments, we cannot investigate the degree to which God intervenes in the physical world.

Many advocates on either of the extremes (CADEs, YECs, OECs) claim that this is a scientifically or theologically indefensible position. Atheistic Darwinian evolutionists claim that one cannot invade the space of unbiased scientific investigation with a god that cannot be investigated. Creationists claim that for Christians to affirm evolution is to deny the truth of an infallible bible and that such an approach leads to a slippery slope ending up in apostasy.

Rather than to favor any one of the versions of TE presented above, my approach will be to describe examples of evolution and postulate how some of these versions of TE might work in these examples. From these examples and the applications of the different versions of TE, the reader can then draw conclusions as to which version seems to work better. The examples that I present here include (1) evolution of birds, with Archaeopteryx as the intermediary; (2) evolution of mammals after the death of the dinosaurs; and (3) evolution of hominids to Homo sapiens.

EVOLUTION OF BIRDS

In either 1860 or 1861, Hermann von Meyer discovered a single feather in a limestone deposit in Solnhofen, Germany, and attributed it to an ancient bird species. Later in 1861, a skeleton was unearthed near Langenaltheim, Germany, indicating a reptile-like species that appeared to be the first-known feathered dinosaur, termed Archaeopteryx.[114] Since that time, eleven more Archaeopteryx skeletons have been discovered in this region of Germany, adding certainty that such a feathered dinosaur

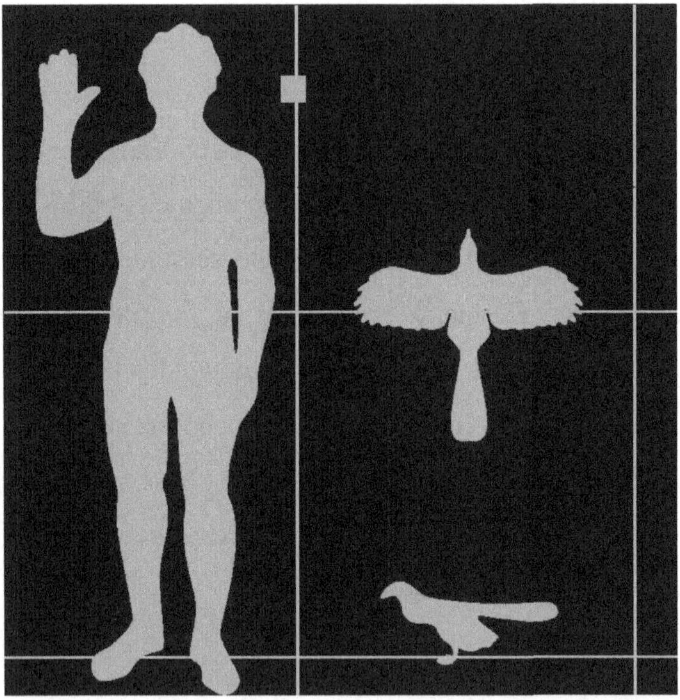

Figure 7.2. Archaeopteryx lithographica morphology and size in relationship to humans (see Matt Martyniuk, "File: Archaeopteryx Scale.svg," Wikimedia Commons, last edited [02-08-2011], [https://commons.wikimedia.org/wiki/File:Archaeopteryx_Scale.svg]).

existed around 150 million years ago. Figure 7.2 shows Archaeopteryx morphology and size in relation to humans. As Charles Darwin stated in his fourth edition of *On the Origin of Species*:

> *Some authors had maintained that the whole class of birds came suddenly into existence during the Eocene period; but we now know, on the authority of Professor Owen, that a bird*

> *certainly lived during the deposition of the upper greensand; and still more recently, that strange bird, the Archaeopteryx, with a long lizard-like tail, bearing a pair of feathers on each joint, and with its wings furnished with two free claws, has been discovered in the oolitic slates of Solnhofen. Hardly any recent discovery shows more forcibly than this how little we as yet know of the former inhabitants of the world.*[115]

While the anatomy of Archaeopteryx has led to questions of whether it was capable of true flight as birds are, it may have had certain limited abilities to glide or jump from one location to another using downstroke-only flapping motions. Archaeopteryx also had features characteristic of other small Mesozoic dinosaurs, including jaws with sharp teeth, three fingers with claws, and a long, bony tail. Taken together, paleontologists suggest, the findings of Archaeopteryx skeletons indicate a bird precursor that was halfway between a nonavian dinosaur and a true bird. Furthermore, this indicates that Archaeopteryx was an evolutionary link between dinosaurs and birds.[116]

If we were to envision the evolution of birds through an Archaeopteryx precursor, the invoking of BioLogos or classical theistic evolution would not involve any special divine intervention. We would expect that Darwinian natural selection would be sufficient to bring about the diversity and success of birds in their environment.

Punctuated interventional TE could suggest some different scenarios on how bird evolution occurred. Feathered dinosaurs such as Archaeopteryx could have arisen through normal variations over time. However, the environment might have changed in a directed way to favor those dinosaurs that had ability to fly. For instance, if a dinosaur can fly, it has an increased range and possibly an increased food supply compared to a dinosaur that moves only on land. A dinosaur that can fly may also have an increased reproductive capacity. We see a number of possibilities here. If the environment remained the same, a flying dinosaur may have had an advantage over land dinosaurs. No intervention would have been required. However, if the environment did change, there might have been an advantage to flying dinosaurs. For instance, if islands are now separated by bodies of water, a flying dinosaur might have an advantage in reaching those islands. The formation of these new environments might be considered a random event, or could be considered to be an intervention. Either way, the flying dinosaur might have an advantage. Punctuated interventional TE cannot distinguish between the random event and the divine intervention. However, it leaves the possibility open that the environment was formed in a nonrandom way and that the new environment favored the new living form—in this case, a dinosaur that could fly.

The Day the Dinosaurs Died

One of the better-known of geological events on planet Earth was referred to as the Cretaceous-Tertiary (K-T) extinction, a mass extinction of more than 75 percent of all life on Earth around 66 million years ago. This included most of the large terrestrial dinosaurs that dominated the Earth at the time. The source of the extinction was traced to a layer of clay with a high level of the element iridium, which is normally rare on Earth. Fossils of all dinosaurs and other extinct flora and fauna were present below the K-T boundary but were not present above the boundary in geologic sediment. It was hypothesized by Luis Alvarez and his son, Walter Alvarez, in 1980 that this K-T boundary was in fact due to the impact of a comet or asteroid on the order of ten to fifteen kilometers in diameter.[117] The impact was thought to have released energy equivalent to several million nuclear weapons and to have created a massive dust cloud throughout the Earth. This asteroid or comet was hypothesized to have landed in Mexico's Yucatan Peninsula in an area referred to as Chicxulub, where a 180 kilometer prehistoric crater was identified.[118] The contemporaneous findings of the impact crater and the worldwide layer of the K-T boundary along with the massive extinction strongly suggested that the massive extinction was caused to a large degree by the impact of the comet or asteroid. Interestingly, although so many species were obliterated in this event, it provided an opportunity for new species to evolve, including a wide variety of mammals (sea mammals, primates, pigs, horses) as well as birds, reptiles, boney fishes, and other chordate sea creatures.

This K-T event and the subsequent biological evolution could be viewed through the glasses of classical theistic evolution, or BioLogos, as traditional Darwinian evolution. Variations of surviving species were hypothesized to compete for available ecologic niches, and survival of the fittest led to evolution of new species to fill those niches. No special divine intervention would be needed to bring about this new evolution.

We can also view the K-T event and subsequent biological evolution from the point of view of punctuated interventional theistic evolution. The event of the comet or asteroid impact can be regarded as either a random event or a divine intervention. Again, we cannot distinguish scientifically between the random and the divine intervention. Either is possible. However, we cannot rule out that a divine intervention—in this case, the impact of a comet or asteroid—resulted in creation of new ecological niches and in the evolution of new species through the processes of Darwinian survival of the fittest. However, it is still remarkable that the K-T event was not of a sufficient magnitude to obliterate all life, but was just large enough to kill off 75 percent of the existing life forms. Thus, the Earth was repopulated by new species that were successful in their new environments. Again, it is not possible to distinguish between a random impact and a directed intervention. The result is the same as in classical TE, but the possibility exists for a directed evolution through an intervention favoring the impact of a comet or asteroid with the earth.

Evolution of Hominids

Hominids include existing primate species of Homo sapiens, bonobo, chimpanzee, gorilla, and the orangutan as well as their extinct precursors. Evolutionarily, humans diverged from their closest living relatives, the chimpanzee and bonobo, 6 million years ago (see figure 7.3). Gorillas diverged from humans, chimps, and bonobos approximately 8 million years ago, and precursors of orangutans split off from other hominids approximately 18 million years ago.

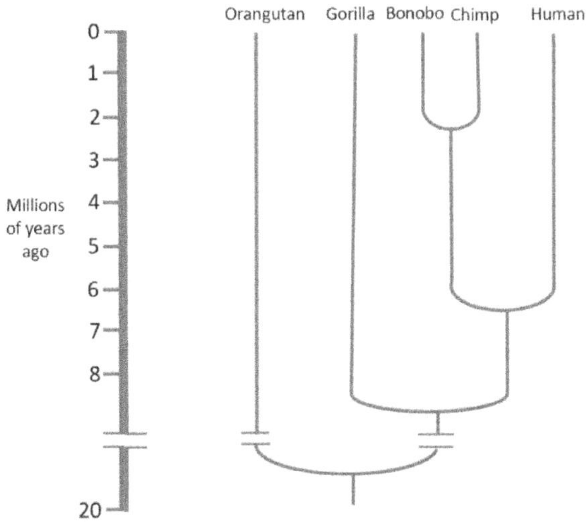

Figure 7.3. Hominid evolution in the last 20 million years (adapted from B. MacEvoy, "Human Evolution," handprint.com, last revised [08-08-2014], [https://www.handprint.com/LS/ANC/evol.html]).

In the last fifty years, great progress has been made in deciphering the evolution of humans over the last approximately 5 million years from a chimp-like precursor. Figure 7.4 shows the lineage of human precursors from 5 million years ago until the present. It is noteworthy that this diagram looks more like a bush than a tree because of the number of different hominid precursors existing simultaneously throughout the African continent. In 1974 Dr. Donald Johanson, a paleoanthropologist from Case Western Reserve University, and his graduate student Tom Gray were searching a site in the Middle Awash of East Africa and came across a trove of fossils that turned out to be derived from a single individual hominid that lived approximately 3.2 million years ago.[119] This hominid, which became known as "Lucy," had many of the characteristics of its relatives the chimps and bonobos but had pelvic and lower leg anatomy to suggest that it walked upright. Johanson and fellow anthropologist Tim White from the University of California at Berkley found that this skeleton was from a species common to thirteen other fossil skeletons found in the area. This species became known as Australopithecus afarensis. Furthermore, the fossils were connected with fossilized footprints discovered by Mary Leakey in Laetoli, Tanzania. These footprints suggested that Lucy was in fact one of the first hominids with the ability to walk bipedally. Lucy, while a small hominid (1.1 m [3 ft, 7 in] tall) with a relatively small brain (400–550 cc) had developed the anatomy to walk bipedally on the ground as well as to use its arms to climb trees in an arboreal environment.

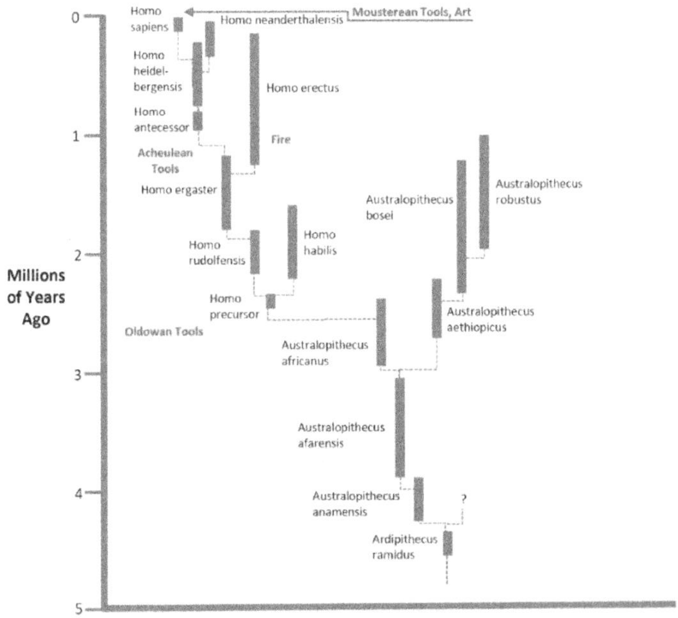

Figure 7.4. Hominid evolution from 5 million years ago to present (adapted from data of D. Johanson and B. Edgar, From Lucy to Language [New York: Simon & Schuster, 1996]).

In 1994, Tim White and Japanese paleoanthropologist Gen Suwa discovered another fossil skeleton from a species that preceded Australopithecus afarensis and was dated to 4.4 million years before the present. This species finally became known as Ardipithecus ramidus, because of significant anatomical differences that distinguished it from Australopithecus.[120] In 2001, another fossilized skeleton was found in the Middle Awash that dated back to 6.5–5.5 million years before present. This subspecies became known as Ardipithecus ramidus kadabba. These ancient species of A. ramidus had pelvic and lower leg anatomy to suggest that individuals walked upright but had an opposable toe that indicated that they were walking on tree limbs as well as on the ground. Although the connections to other forms have been controversial, current information indicates that Ardipithecus is on the line leading to early forms of Australopithecus, including A. afarensis. Clearly, the Australopithecus line exploited the bipedal ability, but it is currently unclear what that advantage might have been. It is possible that in mixed forest and savanna environments, that ability to walk on the ground and to move out of the trees into the plains was an advantage. Walking bipedally may also have been advantageous in that it freed up the arms and hands for other activities.

At approximately 3 million years before present, there was a split of A. afarensis to A. africanus and Australopithecus aethiopicus in eastern Africa. At 2.5

million years before present, the genus Homo branched off of the A. africanus line. Hominids of the genus Homo were characterized by a larger brain case (> 550 cc), greater manual dexterity with an opposable thumb, a more upright stance, and the ability to walk/run long distances. Anthropologist Elizabeth Vrba of Yale University has hypothesized that climatic changes at 2.5 million years ago accelerated hominin evolution as a shift from the more wet forest environment to the drier savannas in East Africa gave an advantage to species that could walk upright and compete in the predominantly savanna environment.[121] Successful species in the Pleistocene (starting 2 million years before present) included Homo ergaster, Homo erectus, and Homo habilis. The combined increase in manual dexterity and neural function led to first the development of the Oldowan tool industry, consisting of simple flaked stone tools, and later on the Acheulean tool industry, consisting of more sophisticated hand axes made of flaked flint and other stones. Homo erectus in particular expanded its range out of Africa to areas of China and Java. Homo erectus was also able to make use of fire for cooking food. The advanced mobility and tool-making skills made possible the development of a hunter-gatherer lifestyle. Because Homo erectus had developed dermal sweat glands, they could run long distances without tiring, making possible what is known as persistence hunting. In this form of hunting, the animal hunted could only run so far without stopping to sweat by panting. The hominid could run much farther, partly due to dermal sweating. At this point, the exhausted animal had lost the race and was killed at close range. Also, because Homo genus members could cover large distances, this ability allowed them to gather more amounts of available vegetation for nutrition.

Over the last approximately 1 million years as an offshoot of either Homo erectus or Homo ergaster, larger-brained hominid species developed, including Homo antecessor, Homo heidelbergensis, Homo neanderthalensis, and anatomically modern Homo sapiens. Homo neanderthalensis was particularly adapted to colder climates and migrated into areas of the Middle East and Europe. Neanderthals were present in Europe and East Asia along with the related Denisovans for more than three hundred thousand years, but at low populations of no more than fifteen thousand individuals for all the subcontinents. They had relatively primitive tool-making skills, ability to use fire, and could use animal skins for protection from the cold. Homo sapiens moved out of Africa into Europe approximately sixty thousand years before the present. They had better tool-making skills for hunting and practiced other industries including the use of needles for tailored clothing from animal hides. Because the Neanderthals and homo sapiens were close cousins, there was occasional interbreeding of the two populations, as indicated by a small amount of Neanderthal DNA that persists in present-day human populations. By all archaeological evidence, the Neanderthals went extinct approximately thirty thousand years ago. It is unclear whether they were out-competed by the humans, were preyed upon by them, or were simply absorbed into the Homo sapiens population. Much work has gone into examining the abilities of Neanderthals in comparison to those of fully modern humans. Much research has also been done to analyze the significance of the contribution of the Neanderthal genome to the present-day human genome.[122] This transition has certainly been an interesting page in the evolution of the human race. Over the subsequent thirty thousand years the human population expanded into Asia, Southeast Asia, and the Americas to be the dominant hominid species on the planet.

This was an evolutionary scenario where a number of hominid life forms existed simultaneously and eventually a few, or even one, dominated the environment. This can be consistent with a number of the variants of theistic evolution. In hominid evolution, much of the bipedal ability and the development of advanced skills can be seen as an adaptation to the environment and survival of the fittest according to Darwin's theory. However, it is difficult to assign survival value to such human qualities as appreciation of art, appreciation of good food, and spiritual values including a belief in a transcendent god, which are characteristic of human behavior. Within a short time, anatomically modern humans throughout Europe showed abilities in producing cave art and took a big interest in burial practices for their dead. They also had highly developed social skills that led to the development of written and spoken languages, organization of cities, and development of a variety of technologies. All these skills were developed after humans were the dominant species on the planet. Thus, it is difficult to see how developing these skills would lead to a survival benefit in the Darwinian sense.

DISTINGUISHING CLASSICAL TE AND PUNCTUATED INTERVENTIONAL TE

All of the examples of theistic evolution to a certain degree leave open the issue of what is a better fit in terms of what we see in nature: classical TE or punctuated interventional TE. Two questions come up as we distinguish between classical TE or BioLogos and punctuated interventional TE:

> *(1) Would the mechanism of Darwinian evolution invoked in classical TE have been sufficient to bring about the evolution and creation of living forms as we see them in nature, independent of any divine intervention?*

This question can best be answered by example. What we see in the evolution of Darwin's finches is what you would expect from Darwin's theory. The finch species adapted to the different environments by generating some variant individuals that were fitter in their respective environments than others, and the fittest survived the best. We see this in hominid evolution with the development of bipedalism as well. Hominids that were bipedal were in many ways better suited to a grasslands environment as opposed to an arboreal environment. Walking or running had a clear survival advantage for hunting and food gathering. For all of these anatomical developments, we need not invoke any special divine intervention to accomplish these changes. However, this is not to say that God did not guide the processes in some way, nor is it to deny that he was upholding the natural principles in such a way that they would harmonize with the rest of his creation. Intelligent design theorists have questioned as to whether a complex organ such as the eye could have originated by a stepwise, evolutionary process. However, even with the eye, through vertebrate evolution, we see the possibility for change over millions of years, consistent with stepwise evolution, leading up to a complex, camera-like eye and a brain that can interpret those images. The classical TE advocate argues that complex evolutionary developments originated solely from natural Darwinian principles operating without any need for supernatural intervention.

> *(2) By invoking punctuated interventional TE, does one invoke an intervention that biases the outcome of what would otherwise be seen as a random outcome, and is such a supposed intervention therefore logically inconsistent with classical Darwinian evolution?*

The argument of the punctuated interventional TE advocate originates from the nature of the transcendental God, that he is both omniscient and omnipotent in a realm that extends beyond the physical universe. In this respect, it is posited that he not only establishes and upholds the physical laws but also is able, on rare occasion, to bias conditions so that they favor a particular outcome. The purely Darwinian advocate then objects that this opens a can of illogical worms that invalidates the normal occurrence of survival of the fittest determining the outcome. This would be true if the hand of God in such an intervention were in any way evident in the physical world. For instance, in the K-T extinction, if the hand of God were in any way evident in favoring the collision of a comet or asteroid with the Earth, it could be argued that Darwinian principles were suspended, and therefore invalidated in that instance. However, the argument of the punctuated interventional

TE advocate is that God's reserving the right to intervene in the physical world is always at a level that would not be seen in the physical world. Furthermore, the instances of intervention would always be of such a low incidence that they could never be detected statistically. The experimental observances would always appear according to random probabilities. Thus, we can still hold to God's nature of omniscience/omnipotence, and at the same time the carrying out of the experimental method would never be impaired by such exceedingly rare interventions.

Other Criticisms of Theistic Evolution vs. Young and Old Earth Creationism

Many young earth and old earth creationists have questioned whether theistic evolution as proposed here could account for the complexity of species observed in our world. While Darwinian evolution through survival of the fittest has been invoked as the mechanism for the variation in living forms that is observed, creationists have questioned whether this could account for macroevolution, which leads to development of significantly different physiological structures. Creationists have also questioned whether Darwinian evolution could account for the development of primordial macromolecules such as DNA, RNA, and proteins from simple building blocks by purely chemical means. It is generally argued by creationists that random chemical reactions and random mutations of genetic material would not be sufficient for the development of these complex sequences of nucleic acids and proteins.

It should be pointed out at the outset that Darwinian evolution as formulated by Darwin was not meant to explain the origin of life from nonliving components. It only can explain the development of new and novel species from a common precursor. So even if development of initial life forms such as unicellular organisms and subcellular components is either improbable or unimaginable given the known physical and chemical interactions of macromolecules, this is outside the realm of what the theory of Darwinian evolution proposes. Future investigation may reveal ways in which these structures might have developed. However, these discussions are outside the realm of what Darwinian evolution attempts to explain. The discussion is still open about whether Darwinian evolution can explain macroevolution of new and novel structures in the species that evolve. Examples have been given for avian and hominid evolution. It is also proposed here as a principle of theistic evolution that while "random" change predominates in Darwinian evolution, God is involved in some way in the changes that occur over time in the evolutionary processes. As has been stated previously, these interventions are sufficiently rare as not to interfere with the rigorousness of the scientific method. Furthermore, because God is transcendent, we cannot investigate experimentally the way that he is involved in maintaining his creation. The BioLogos and

Classical versions of TE posit that we do not have to invoke any special divine interventions to account for the development of new and novel structures and species. While God's involvement in his creation remains an open question, these versions of TE propose that evolutionary processes on their own will lead to new and novel physiological structures leading to macroevolution.

∞

All the variants of theistic evolution, regardless of the degree of intervention in the creation process (with the exception of modernist TE), assume the existence of a transcendent god who has an inherent care for his creation. This would be in contrast to Darwinian evolution, which does not assume any existence of a god as essential for the creation process. Theistic evolutionists agree that God created, but they maintain that he used the principles of Darwinian evolution to carry out that process. Many Darwinian evolutionists or creationists on the extremes of the spectrum may argue that the theistic evolutionist position is a compromise position meant to satisfy the conflicts that are seen in this debate of "creation versus evolution." Rather, the theistic evolutionist position can be regarded as the best approximation of what we see in nature. We see strong information that supports the evolution of living forms through survival of the fittest. However, we also see reflected in nature a god who is transcendent and has created in a particular direction that has led to evolution of humans who could appreciate his being. This is another way of saying that we were created by God in his image.

The degree to which such a god would intervene or direct such a process is unknown, but a wide range of interventions are certainly possible based on his omnipotence and omniscience. We assign these qualities to a transcendent god, so we have to assume that he would act in accordance with these qualities in directing creation. So, the punctuated interventional TE model remains within the realm of possibility, as do BioLogos and classical TE. If we were to extend our understanding of God into the Christocentric theology of orthodox TE, we might also link the creative god with the God who also suffered for us in the death of Christ. In Old Testament theology we read in Genesis 1 that God created man in his own image. Thus, theistic evolution may indicate that many of our aesthetic and spiritual qualities come directly from God as being imprinted from his nature, and without any inherent survival value in the Darwinian sense. However, these theistic approaches to biological evolution remain consistent with and harmonious with application of the scientific method. In assuming the involvement of a transcendent god in creation, we do not take away from the application of the highest level of scientific rigor in applying the scientific method. Moreover, the application of theistic evolution allows us to appreciate the majesty of God's creation and at the same time maintain a connection to the strong experimental evidence of Darwinian evolution.

Chapter 8:

Summary

To summarize our discussions, an argument is made for the existence of a transcendent God who created the world and created man in his own image. This is contrasted with the atheistic view of the natural world, which hypothesizes the absence of this transcendent God. Furthermore, the atheistic view hypothesizes that man has in his own thinking invented the concept of an all-powerful god. However, the theist point of view proposes a transcendent God who has an existence independent of man and exists outside of the space-time universe in which we live. While the acts of God are consistent with the biblical narrative, we do not invoke a literal interpretation of the texts, nor do we view the Bible as stating explicitly how physical events of creation took place over time and space. This interpretation of the Bible avoids many of the potential conflicts indicating possible differences between what the Bible teaches and what science teaches.

It is hypothesized in this essay that natural laws derived from application of the scientific method are the most reliable way to determine how the universe operates. However, it is God in his sovereignty and omnipotence who governs those laws. In very rare or exceptional events that we would regard as interventions of God, or miracles, such laws are suspended. However, in the conducting of science, we see only the regular patterns of nature and phenomena that allow the derivation of the natural laws that God has established. Hence, it is not possible to investigate scientifically these interventions of the transcendental God. Furthermore, the rare intervention of God in his creation does not prevent us from rigorously studying natural laws through the scientific method. Oftentimes, by doing observational science close up, we fail to see the work of God in his governing of the natural world. However, if we observe at a distance and with a wider perspective, we are more likely to see a reflection of the magnificent work of God in his creation.

In this essay I have reviewed a number of life science areas in which we see a reflection of God's glory in creation. In structural biology we see detailed structures at the molecular scale that carry out precise functions in cellular metabolism. We see this further in the structure of DNA, in replication of DNA, and in functions leading to protein synthesis. In the study of the genome, we see even further, discerning the language of God spelled out in sequences that result in whole organisms, including humans made in his image.

Other areas that bear the mark of God on creation include enzymology and cellular signaling. In these areas, we see the remarkable interplay of protein macromolecules that are able to carry out cellular metabolism. We see cellular communication at the physiological level that makes it possible for organisms to control their microenvironment through extracellular hormones that signal to the inner workings of the cell. Again, the detail and complexity are reflections of God's majesty in the creation but carried out through particular means that include the evolutionary process.

In this essay, I have presented a range of positions that people hold concerning the processes of creation and evolution, including Darwinian evolution, theistic evolution, intelligent design, and old earth/new earth creationism. In emphasizing theistic evolution, I intend neither to find a compromise between de novo creation and atheistic evolution nor to minimize the role of God in creation. The intent is to simply seek the best fit for how creation would have happened given what we see in nature. We see evidence for a universe that was created approximately fourteen billion years ago. We see an origin of unicellular forms of life approximately four billion years ago. We then see in the archaeological record evidence of successive forms of life of increasing complexity leading to the physiological complexity that we see in mammals as recent as one hundred million years ago and eventually humans in the time period of one hundred thousand years ago. All of these changes in life forms can be explained through Darwinian evolution.

However, when we introduce the possibility of a transcendental God into the picture, we have to ask how this God has been involved in creation. Did he simply intervene in the initial big bang event of creating the universe, or does he continue to intervene sporadically and rarely, as the punctuated interventional theistic evolutionists would suggest? This essay does not seek to answer this question, but only indicates the possibility that God has worked in the evolutionary process using primarily Darwinian means of evolution and that God can nevertheless intervene. With this outlook, we can remain true to what the scientific method tells us about the nature of the world. We can remain true to a rigorous and logical understanding of natural science. We do not sacrifice any of the ability to rigorously study nature or to draw conclusions about what we find. If anything, we come to a greater appreciation for the world as we carry out our investigations.

Chapter 9:

Epilogue

In April 2020, the first wave of the COVID-19 pandemic had just started to take hold. I had been having some persistent indigestion that I attributed to having eaten some bad food, and finally consulted my primary care doctor. He suggested some antacids, which had little effect. I came back to him in a few weeks also with a report of unusually dark urine. To be safe, he then ordered some blood tests, which showed abnormal liver function. I went for an abdominal ultrasound. Afterward my primary care doctor called with the words that no one wants to hear: “It could be cancer.” Normally, my primary care doctor is conservative about anything that he says, so I took his words particularly seriously. He didn’t offer much detail, except that there were spots on my liver. A subsequent talk with a gastrointestinal specialist revealed that this was probable pancreatic cancer.

I was scheduled for an endoscopic retrograde cholangiopancreatography (ERCP) procedure to directly look at what was going on with my pancreas as well as to take some samples of liver sites. Results conclusively showed pancreatic cancer but were inconclusive as to the spread of cancer to the liver. A laparoscopic surgery was planned to sample these sites more comprehensively. Absence of spread to the liver would indicate a much better prognosis, with possible surgery to remove the primary tumor. Spread to the liver would rule out any possible surgery and leave palliative chemotherapy as the only option. While I certainly hoped that this procedure would show lack of spread to the liver, I was prepared for any outcome. As was confirmed by the laparoscopic surgery, the cancer had spread to the liver and was categorized as stage 4. This is the most common outcome for pancreatic cancer patients. Eighty percent of pancreatic cancer patients are diagnosed initially with stage 4 cancer that has spread to the liver. This is why pancreatic cancer is so deadly, because in most cases, by the time it has been diagnosed, it has already spread to surrounding organs such as the liver.

Needless to say, I was in shock for the first few weeks after I had gotten this diagnosis. At sixty-six years old, I had experienced minimal health issues in my life, had low risk for GI cancers, and had a minimal history of cancer in my family. It was like being struck by lightning. Even with the low probability, one can only acknowledge that it has happened. In spite of the poor prognosis, I chose to believe that there would be some new therapies out there that could give a better result than the standard chemotherapies. Also, as a theist and a Christian, I had the conviction that God was able to heal me of this affliction should it be his will.

For me, getting this diagnosis and confronting my mortality is particularly relevant as a physical and spiritual reality. It has forced me to put off all the trivial aspects of this world and confront both the physical aspects of this disease and the spiritual aspects of the world in which I now find myself. I should say initially that after getting this diagnosis, I did not feel that God had abandoned me or treated me unfairly. On the contrary, I felt in many ways closer to God, and I have felt that God is more present, and able to meet my needs in this situation. Rather than give a blow-by-blow account of what has happened to me on this journey, I want to ask and answer some basic questions that come to mind when one deals with this kind of a diagnosis in view of the physical and spiritual realities. The answers that I will give are from the point of view of a theist and a Christian. It is true that a nontheist or someone from a different spiritual tradition might have different answers to these questions, and I definitely respect these differing views. The point is to pose various questions for someone who is faced with the basic challenges of dealing with his or her own mortality. The questions range from our own technological/medical means to God's power and sovereignty in dealing with our situation. The intent is to aid anyone in navigating the uncharted waters of a potentially fatal diagnosis.

(1) How have things changed in your life after this diagnosis?

Since being diagnosed with pancreatic cancer fifteen months ago, I have suffered few effects of the actual cancer. Most of the ill effects have been due to fourteen months of chemotherapy. This included a one-month stay in the hospital for treatment of pneumonia, sepsis, and some internal bleeding, all brought about by side effects of the first few rounds of FOLFIRINOX chemotherapy. A subsequent treatment with FOLFOX had lesser side effects but proved not to have any effects on the progression of the cancer. For the last ten months, I have been on a regimen of Abraxane and Gemcitabine chemotherapy. This treatment has so far been more effective in stabilizing the cancer, and at the same time has had more tolerable side effects. These side effects range from occasional fatigue to gastrointestinal irregularity and low blood cell counts.

While I have not suffered major physical effects, my regular day-to-day routine has totally changed. Whereas I had been retired and working part-time as a medical writer, I now found that it made more sense to be fully retired and to concentrate on my recovery. Also, with such a diagnosis, I quickly realized that planning for a twenty-year retirement was not in the cards. Pancreatic cancer patients have an overall survival of 5 percent after one year. While being adamant about not wanting to dwell on the statistics, I soon realized that long-term survival would mean years and not decades. All those plans for an extended retirement, travel, and a relaxed life in the seventies and eighties went quickly down the drain.

As with others that suffer from advanced cancer, for me the future is exceedingly uncertain. However, I have found a greater spiritual presence of God to assure me that his support is there in both the physical and the spiritual realms, whatever the physical limitations of this life. For all the things that have changed after this diagnosis, the one thing that hasn't changed is God's presence to guide me and show me the way through the changed circumstances. Even with the heightened presence of God in my life, it continues to be a major challenge to accept my shortened life span. For most of us, we have this sense of immortality—that death is so far out there that we need not even think of it. This is for the most part an illusion because we all have finite life spans, subject to the shortcomings and defects of our own bodies. We all, if we were honest, would have to acknowledge that there are physical events or physical illnesses that will result in our eventual demise. The advantage for the Christian believer, like myself, is that I don't have the anxiety or insecurity that comes from anticipating some unplanned or unanticipated event. I know that God has this fully under control and that he has a place prepared for me when that time comes.

(2) Can God heal someone completely, given this diagnosis?

Shortly after I was diagnosed with pancreatic cancer, I had my pastor come over to my house along with another deacon to pray for my healing. We agreed that not only is God able to do such a healing in a miraculous way, but we could expect him to do it if it was his will. Many would find such an attitude to be self-contradictory. If you really believed that God would do this miraculous healing, why would you provide him a way out by saying that if it didn't happen, therefore it just must not have been his will? Wouldn't that be a way of making accommodation for a god that was not able to do this healing in the first place? However, the God that I have come to know way back in my earlier days is a transcendent and all-powerful God, capable of intervening in a multitude of human affairs, including the healing of an incurable disease such as cancer. Such a God has also been able to maintain physical laws and to at the same time allow for consistent scientific inquiry through the scientific method. So, for one who has had this view of a transcendent God, it is possible to believe that he would consider physical healing in the realm of what is possible or even likely. That he has not yet done that does not affect my trust in him or decrease my confidence that he will do it in the future. My view of him has been enlarged to include an all-knowing and all-powerful God that may have plans that go far beyond what I can envision. Jesus, in his time on earth, was noted in the Gospels to have healed many people. Many people have hypothesized that such healing was only for the apostolic period and was not to continue into the present time. However, the scriptures don't appear to put any temporal limit on when healing is available. In the book of James, we are encouraged that "the prayer of faith will restore the one who is sick, and the Lord will raise him up" (James 5:15). So, we can expect that miraculous healing by God is still available in our time, but only as a part of his larger will, which may be far more complicated than what we as humans can envision.

(3) How might modern medicine and innovation be God's way of bringing about a healing in this situation?

There is a good proportion of people who when faced with what looks like a life-threatening disease will refuse to go to a doctor for medical treatment. This is not to be confused with people whose disease or condition is so advanced that medical treatment might prolong life with no quality or those that have a living will to prevent heroic intervention at the end of life. There are, however, many people for whom there is reasonable treatment but who, for one reason or other, don't seek out information on that treatment, or choose not to go down the road of pursuing that treatment. Usually, the result is an unnecessarily shortened life because of the rapid progression of what might have been a treatable disease.

Given that it might be possible to extend life with good quality, one might first ask why it is of benefit to do so. The reasons are many, including enriching the lives of family and friends who value us and, in general, making the most of each day that we are given. In Psalm 90:12 the psalmist speaks a prayer that can serve as an admonition to us: "So teach us to number our days, that we may present to you a heart of wisdom." So, extending our time on this earth has not only the benefit of giving us extra time with others but also that of helping us grow in wisdom that is appreciated by God. As much as we might be content to go and be with God, God may yet have lessons for us to learn or teach to others on this earth.

I would conjecture that it is through such available medical care that God often works in bringing about healing. In Hebrews 12:12–13 we have the following statement: "Therefore, strengthen the hands that are weak and the knees that are feeble, and make straight paths for your feet, so that the limb which is impaired may not be dislocated, but rather be healed." So according to this verse, healing by God might be brought about by an active process of seeking out and following the appropriate medical care. One of the unique qualities of humans is that we put a high value on extending life by available means and improving the quality of life. There have always been a small fraction of people advocating euthanasia for those suffering from incurable or debilitating diseases. However, the medical community and the various religious communities have mainly emphasized extending life wherever possible and improving quality of life wherever possible.

Another conjecture that I would make is that being a part of clinical trials and experimental medicine, for the treatment of a disease that is as yet untreatable by standard methods, may also be a way that God advances the treatment of disease and the development of new therapies. These clinical trials may be another part of God's plan to bring about healing. Not only is it possible that we, as participants, will benefit by such experimental therapies, but it is possible that others in the future will benefit from these clinical trials. Consider, for instance, childhood cancers such as leukemia that were incurable sixty years ago but are now considered highly curable. Not only have the children cured of these cancers gone on to live full lives, but they have contributed immeasurably to our society. So, there is no doubt that God has used experimental medicine to benefit society and to bring about further healing of those afflicted with what might be currently incurable diseases. For those of us suffering from such diseases, it's worth considering whether clinical trials may be worthwhile. In my own case, because pancreatic cancer has such a poor prognosis, and because available therapies are so limited, it has been worth it for me to consider possible clinical trials. It is an area where new therapies are on the horizon, and participation in the right trial could have a benefit for me and for those that will come down with this disease in the future.

(4) How does God's spiritual reality address our mortality, and what hope does He give us in light of the ultimate result of a life-threatening illness?

With respect to a life-threatening illness, Jesus gives us a particular encouragement as expressed in the fourteenth chapter of the Gospel of John:

> *Do not let your heart be troubled; believe in God, believe also in Me. In My Father's house are many rooms; if that were not so, I would have told you, because I am going there to prepare a place for you. And if I go and prepare a place for you, I am coming again and will take you to Myself, so that where I am, there you also will be. (John 14:1–3)*

For those facing death, Jesus clearly states that there is an afterlife and that those who believe in him will be welcomed into an afterlife of never-ending bliss. On a number of occasions, Jesus makes clear that this is something that we can depend on. So, as much as he cares for us in this physical life—and he has many plans for us in this life—the end result on the other side of the grave is a life with him in the spiritual realm. Many will ask the questions: "What if I don't believe in Jesus? Is there any inheritance for me in this eternal life?" It is not my purpose here to make judgements about these cases. My understanding of the transcendental God who rules over all things and knows all things is not that good. It is only to state that we have a clear path made possible by the promises of Jesus. Of all the promises that we have from Jesus, this is one of the surest promises—that of eternal life beyond the grave. We as followers of Jesus ultimately follow him to the cross, to our own death, and then follow him into the resurrected life, as he has been a forerunner and has gone before us to specifically prepare the way for us. Jesus has some purpose for us suffering this particular type of disease. He may, as told in the previous question, prolong our time through medical technology. He may have particular purposes for how we live our lives. However, ultimately, we all face the same curse of mortality. It is the same for all, some sooner and some later. Some will suffer a painful death. Others will not suffer to the same extent. Some will suffer a sudden, unexpected death, and others will go through a longer process. However, his teaching is clear that for those who believe in him, he prepares for us a place of never-ending bliss, a place that is our true home. It is clear that this present physical life is not our true home. It is only the prelude of something greater and more permanent to come. The author of the book of Hebrews in chapter 11 states of the heroes of the faith that "they were strangers and exiles on the earth." Furthermore, they desire a better *country*, that is, a heavenly one. Therefore, God is not ashamed to be called their God; for He has prepared a city for them. (Hebrews 11:16)

Again, we have the strong assurance that God has prepared a place for his chosen ones after death.

Even stronger than this is Paul's defense of the resurrection of the dead in 1 Corinthians 15, where he states:

> *If there is no resurrection of the dead, then not even Christ has been raised; and if Christ has not been raised, then our preaching is in vain, your faith also is in vain. Moreover, we are even found to be false witnesses of God, because we testified against God that He raised Christ,*

> *whom He did not raise, if in fact the dead are not raised. For if the dead are not raised, not even Christ has been raised; and if Christ has not been raised, your faith is worthless; you are still in your sins. (1 Corinthians 15:13–17)*

So, this hope that we have in the resurrection and our ultimate fate in going to be with Christ after death is one of the surest promises that we have in the New Testament. As is also stated in Hebrews 6:

> *This hope we have as an anchor of the soul, a hope both sure and reliable and one which enters within the veil. (Hebrews 6:19)*

So, as a spiritual reality, we have to consider these biblical truths, written so long ago, as touching on a matter so apparent to the writers of scripture that they agreed wholeheartedly on this issue of resurrection and the life everlasting. The vividness of these truths spoken in a time very different from ours speaks to us as well, even though we have such a greater understanding of physical truths as reflections of the physical reality that has become apparent in the last few hundred years. Our different times have led to different emphases of what we consider significant. In New Testament times people were looking for a Messiah that would lead them into a new political time, but they discovered in Jesus a new spiritual reality that far exceeded what they had experienced in the physical reality.

∞

Overall, in light of what has been discussed in this essay, we are compelled to consider both the spiritual and physical realities of our world. In our present time, we still have a remnant of the influence of the spiritual reality conveyed in the traditions of Jesus. We have to consider seriously his claims of a divine nature, his teachings, and his claim to a better plan for our lives. However, the dominance of physical truths brought about by the scientific revolution has overwhelmed our thinking in terms of what we see as the apparent reality. We see the molecular reality and the words written in DNA and protein sequences as being more significant than the truths written in scriptural texts. And yet, after all these years, we cannot deny the spiritual reality of both the scriptural texts and the molecular sequences. As we look into these molecular sequences and protein structures, we see a reflection of the greater magnitude and majesty of the spiritual truths that we will only fully appreciate when we get to the other side of this earthly life. In concluding our journey into the physical and spiritual realms, I leave you with these words from Psalm 72 as David speaks about the transcendental God:

> *Blessed be the LORD God, the God of Israel,*
> *Who alone works wonders.*
> *And blessed be His glorious name forever;*
> *And may the whole earth be filled with His glory.*
> *Amen and Amen.*
>
> Psalm 72:18–19

Acknowledgements

I would like to acknowledge numerous teachers and professors that instilled in me an appreciation for the life sciences and set examples for how investigation in the life sciences is done. These include Warner Love, Biophysics professor, Johns Hopkins University, Maurice Bessman, Biology professor, Johns Hopkins University, Richard Cone, Biophysics professor, Johns Hopkins University, Evangelos Moudrianakis, Biology professor, Johns Hopkins University, and Saul Roseman, Biology professor, Johns Hopkins University.

I would also like to make a special acknowledgement to my thesis advisor at the University of Tennessee, Professor Jayant Joshi. Dr. Joshi was instrumental in directing me to the area of cellular signaling and gave me a firm grounding in the area of experimental enzymology.

In preparation of this essay, I would like to thank many who provided comments, including Hitesh Jindal, Gene Forrest, my sister, Beverly Price Rollins, and Bob Banning who provided copy editing for the manuscript. A special thanks goes to Jolinda Pizzirani at Summerland Publishing for help with formatting and publishing.

Lastly, I want to thank especially my family, including my wife Nikki, and sons Glenn and Harry for putting up with me through the writing of this essay. I thank my son Glenn especially for encouraging me to see this essay through to completion and for encouraging me to publish it.

Index

B

C

D

E

F

G

H

I

J

K

L

M

N

O

Q

R

S

T

U

V

W

X

Y

Z

α

β

Endnotes

[1] Francis S. Collins, *The Language of God: A Scientist Presents Evidence for Belief* (New York: Free Press, 2006), 5–6.

[2] C. S. Lewis, *Miracles* (New York: Macmillan, 1960), 30.

[3] All scripture quotations are taken from the (NASB®) *New American Standard Bible®* unless otherwise noted, Copyright © 1960, 1971, 1977, 1995, 2020 by The Lockman Foundation. Used by permission. All rights reserved. www.lockman.org.

[4] J. C. Whitcomb, Jr. and H. M. Morris, *The Genesis Flood: The Biblical Record and Its Scientific Implications* (Nutley, NJ:The Presbyterian and Reformed Publishing Company, 1977).

[5] C. Seddon, *Humans: From the Beginning; From the First Apes to the First Cities* (London: Glanville Publications, 2014), 147, ebook.

[6] R. Dunbar, "The Social Brain and the Cultural Explosion of Human Revolution," in *Rethinking the Human Revolution*, ed. Paul Mellars, Katie Boyle, Ofer Bar-Yosef, and Chris Stringer (Cambridge: McDonald Institute, 2007), 91–105.

[7] C. Seddon, *Humans: From the Beginning; From the First Apes to the First Cities* (London: Glanville Publications, 2014), 149, ebook.

[8] B. Greene, *The Hidden Reality: Parallel Universes and the Deep Laws of the Cosmos* (New York: Knopf, 2011).

[9] Andrew Chaikin, *A Man on the Moon: The Voyages of the* Apollo *Astronauts* (New York: Penguin Group, 1994), 56.

[10] Chaikin, *Man on the Moon*, 121.

[11] Chaikin, *Man on the Moon*, 157.

[12] Chaikin, *Man on the Moon*, 437, 558.

[13] Deborah B. Haarsma and Loren D. Haarsma, *Origins: Christian Perspectives on Creation, Evolution, and Intelligent Design* (Grand Rapids, MI: Faith Alive Christian Resources, 2011), 47.

[14] N. F. Gier, "The Three-Story Universe," chapter 13 in *God, Reason, and the Evangelicals* (Lanham, MD: University Press of America, 1987).

[15] Gier, "Three-Story Universe," chapter 13.

[16] See C. C. Ryrie, ed., *New American Standard Ryrie Study Bible* (Chicago: Moody, 1994), 11.

[17] See S. F. Mason, *A History of the Sciences* (New York: Collier Books, 1975), 49.

[18] C. H. F. Peters and E. B. Knobel, *Ptolemy's Catalogue of Stars: A Revision of the "Almagest"* (Washington, DC: Carnegie Institution of Washington, 1915).

[19] See S. F. Mason, *A History of the Sciences* (New York: Collier Books, 1975), 128.

[20] N. Copernicus, *De Revolutionibus, Book I*, trans. J. F. Dobson and S. Brodetsky, *Occasional Notes of the Royal Astronomical Society* 2, no. 10 (1947).

[21] See S. F. Mason, *A History of the Sciences* (New York: Collier Books, 1975), 136.

[22] Galileo Galilei, *Dialogues Concerning Two New Sciences*, trans. H. Crew and A. de Salvio (New York: Macmillan, 1914); Galileo Galilei, *Dialogue Concerning the Two Chief World Systems*, trans. Stillman Drake (Berkeley: University of California Press, 1967).

[23] Galileo Galilei, *Siderius Nuncius, or The Starry Messenger*, trans. Albert Van Helden (Chicago: University of Chicago Press, 1989).

[24] R. E. Dickerson and I. Geis, *The Structure and Action of Proteins* (New York: Harper and Row, 1969).

[25] Max F. Perutz, introduction to Dickerson and Geis, *Structure and Action of Proteins*, vii–viii.

[26] H. M. Berman, J. Westbrook, Z. Feng, G. Gilliland, T. N. Bhat, H. Weissig, I. N. Shindyalov, and P. E. Bourne, "The Protein Data Bank," *Nucleic Acids Research* 28, no. 1 (2000): 235–42.

[27] Michio Kaku, "The Universe in a Nutshell," YouTube video, 42:14, August 15, 2012, available at brainpickings.org,

https://www.brainpickings.org/?s=%22the+universe+in+a+nutshell%22.

[28] See G. M. Caroe, *William Henry Bragg, 1862–1942: Man and Scientist* (Cambridge: Cambridge University Press, 1978).

[29] D. Phillips, "William Lawrence Bragg (31 March 1890–1 July 1971)," *Biographical Memoirs of Fellows of the Royal Society* 25 (1979): 78.

[30] Phillips, "William Lawrence Bragg," 78.

31 Phillips, "William Lawrence Bragg," 81.

[32] Max von Laue, "Concerning the Detection of X-Ray Interferences," *Nobel Lecture*, June 3, 1920, available at The Nobel Prize, https://www.nobelprize.org/prizes/physics/1914/laue/lecture/.

[33] Phillips, "William Lawrence Bragg," 91.

[34] "Lawrence Bragg: Biographical," available at *The Nobel Prize*, https://www.nobelprize.org/prizes/physics/1915/wl-bragg/biographical/.

[35] Phillips, "William Lawrence Bragg," 101–4.

[36] J. Medawar and D. Pyke, *Hitler's Gift: The True Story of the Scientists Expelled by the Nazi Regime* (New York: Arcade, 2012), 108.

[37] Phillips, "William Lawrence Bragg," 117.

[38] This experience is narrated by Patience in Patience Thomson, "A Tribute to W. L. Bragg by His Younger Daughter," *Acta Crystallographica* A69 (2013): 5–7.

[39] Medawar and Pyke, *Hitler's Gift*, 109.

[40] Phillips, "William Lawrence Bragg," 110–11.

[41] Wikipedia, s.v. "Max Perutz," last edited July 16, 2021, 15:21, https://en.wikipedia.org/wiki/Max_Perutz. See especially under the heading "Early Life and Education."

[42] Medawar and Pyke, Hitler's Gift, 110–11.

[43] S. Everts, "Information Overload," *Distillations* 2, no. 2 (2016): 26–33.

[44] M. F. Perutz and H. Lehmann, *Nature* 219 (1968): 902–9.

[45] M. F. Perutz, *Mechanisms of Cooperativity and Allosteric Regulation in Proteins* (Cambridge: Cambridge University Press, 1990).

[46] P. M. Wassarman, *A Place in History: The Biography of John C. Kendrew* (New York: Oxford University Press, 2020), 104, ebook.

[47] Wassarman, *Place in History*, 51, ebook.

[48] Wassarman, *Place in History*, 86, ebook.

[49] Wassarman, *Place in History*, 105, ebook.

[50] Wassarman, *Place in History*, 123, ebook.

[51] Wassarman, *Place in History*, 116–18, ebook.

[52] Wassarman, *Place in History*, 123–24, ebook.

[53] John C. Kendrew, "The Three-Dimensional Structure of a Protein Molecule," *Scientific American* 205 (December 1961): 96–110.

[54] See Wassarman, *Place in History*, 239–40, ebook; K. Patel, "Perutz Rubbishes Popper and Kuhn," *Times Higher Education Supplement*, November 25, 1994.

[55] Perutz, Introduction to Dickerson and Geis, vii–viii.

[56] H. E. Edgerton, Flash! Seeing the Unseen by Ultra High-Speed Photography (Boston: Hale, Cushman & Flint, 1939).

[57] H. E. Edgerton, Back Dive. 1954. MIT Museum, Cambridge, MA. HEE-NC-54001. MIT Museum, Edgerton Digital Collections. Date accessed 09-23-2021. http://edgerton-digital-collections.org.

[58] James D. Watson, *The Double Helix* (New York: Atheneum, 1968).

[59] Oswald T. Avery, Colin M. MacLeod, and Maclyn McCarty, "Studies on the Chemical Nature of the Substance Inducing Transformation of Pneumococcal Types: Induction of Transformation by a Desoxyribonucleic Acid Fraction Isolated from Pneumococcus Type III," *Journal of Experimental Medicine* 79, no. 2 (1944): 137–58.

[60] Alfred D. Hershey and Martha Chase, "Independent Functions of Viral Protein and Nucleic Acid in Growth of Bacteriophage," *Journal of General Physiology* 36, no. 1 (1952): 39–56.

[61] See E. Chargaff, R. Lipshitz, and C. Green, "Composition of the Deoxypentose Nucleic Acids of Four Genera of Sea-Urchin, *Journal of Biological Chemistry* 195, no. 1 (1952): 155–60; E. D. Chargaff, "On the Deoxyribonucleic Acid Content of Sea Urchin Gametes," *Experientia* 8, no. 4 (1952): 143–45.

[62] Watson, *Double Helix*, 105–8.

[63] Watson, *Double Helix*, 120–22.

[64] Watson, *Double Helix*, 123–26.

[65] James D. Watson and Francis H. C. Crick, "The Structure of Deoxyribose Nucleic Acid," *Nature* 171 (1953): 737–38.

[66] Watson, *Double Helix*, 140.

[67] H. J. Matthaei, O. W. Jones, R. G. Matin, and M. W. Nirenberg, "Characteristics and Composition of RNA Coding Units," *P.N.A.S.* 48, no. 4 (1962): 666–77.

[68] M. B. Hoagland, M. L. Stephenson, J. F. Scott, L. I. Hecht, and P. C. Zamecnik, "A Soluble Ribonucleic Acid Intermediate in Protein Synthesis," *Journal of Biological Chemistry* 231, no. 1 (1958): 241–57.

[69] F. Sanger, S. Nicklen, and A. R. Coulson, "DNA Sequencing with Chain-Terminating Inhibitors," *P.N.A.S.* 74, no. 12 (1977): 5463–67.

[70] K. Mullis, F. Faloona, S. Schraft, R. Saiki, G. Horn, and H. Erlich, "Specific Enzymatic Amplification of DNA in Vitro: The Polymerase Chain Reaction," *Cold Spring Harbor Symposia on Quantitative Biology* 51 (1986): 263–73.

[71] Kary B. Mullis, "The Polymerase Chain Reaction," Nobel Lecture, December 8, 1993, available at *The Nobel Prize*, https://www.nobelprize.org/prizes/chemistry/1993/mullis/lecture/.

[72] P. Banerjee, M. F. Ahmad, J. R. Grove, C. Kozlosky, D. J. Price, and J. Avruch, "Molecular Structure of a Major Insulin/Mitogen-Activated 70-Kda S6 Protein Kinase," *P.N.A.S.* 87 (1990): 8550–54.

[73] J. R. Grove, P. Banerjee, A. Balasubramanyam, P. J. Coffer, D. J. Price, J. Avruch, and J. R. Woodgett, "Cloning and Expression of Two Human P70 S6 Kinase Polypeptides Differing Only at Their Amino Termini," *Molecular and Cellular Biology* 11, no. 11 (1991): 5541–50.

[74] R. Dulbecco, "A Turning Point in Cancer Research: Sequencing the Human Genome," *Science* 231, no. 4742 (1986): 1055–56.

[75] Wikipedia, s.v. "Human Genome Project," www.en.wikipedia.org. See especially the section titled "History."

[76] "Frequently Asked Questions and Resources," National Human Genome Research Institute, last updated June 11, 2020, www.genome.gov/human-genome-project/Completion-FAQ.

[77] "Frequently Asked Questions and Resources," National Human Genome Research Institute.

[78] See E. S. Lander, L. M. Linton, B. Birren, et al., "Initial Sequencing and Analysis of the Human Genome," *Nature* 409, no. 6822 (2001): 860–921.

[79] See J. C. Venter, M. D. Adams, E. W. Myers, P. W. Li, R. J. Mural, G. G. Sutton, et al., "The Sequence of the Human Genome," *Science* 291, no. 5507 (2001): 1304–51; J. C. Venter, "A Part of the Human Genome Sequence," *Science* 299, no. 5610 (2003): 1183–84.

[80] See Consortium, International Human Genome Sequence, "Finishing the Euchromatic Sequence of the Human Genome," *Nature* 431, no. 7011 (2004): 931–45.

[81] "The Cancer Genome Atlas Program," National Cancer Institute, www.cancer.gov.

[82] "Reading the Book of Life: White House Remarks on Decoding of Genome," *New York Times*, June 27, 2000, https://www.nytimes.com/2000/06/27/science/reading-the-book-of-life-white-house-remarks-on-decoding-of-genome.html.

[83] For an account of Buchner's research on enzymes and glucose, see E. Buchner and R. Rapp, "Alkoholische Gahrung ohne Hefezellen," *Berichte der Deutschen Chemischen Gesellschaft* 30 (1898): 209–17.

[84] J. B. Sumner, "The Isolation and Crystallization of the Enzyme Urease: Preliminary Paper," *Journal of Biological Chemistry* 69 (1926): 435–44.

[85] J. H. Northrup, *Crystalline Enzymes: The Chemistry of Pepsin, Trypsin, and Bacteriophage*, (New York: Columbia University Press, 1939).

[86] See studies of Fischer (in M. O. Forster, "Emil Fischer Memorial Lecture," *Journal of the Chemical Society, Transactions* 117 [1920]: 1157–201) and Koshland (in R. Scheckman, "The Nine Lives of Daniel E. Koshland Jr. (1920–2007)," *P.N.A.S.* 104, no. 37 [2007]: 14551–52).

[87] See L. Michaelis and M. L. Menten, "Die Kinetik der Invertinwirkung," *Biochemische Zeitschrift* 49 (1913): 333–69.

[88] See chapter 4 for X-ray crystallographic studies.

[89] See J. M. Orten and O. W. Neuhaus, *Human Biochemistry* (St. Louis: C.V. Mosby, 1982), 321–22.

[90] See E. G. Krebs, "An Accidental Biochemist," *Annual Review of Biochemistry* 67 (1998): xii–xxxii.

[91] Orten and Neuhaus, *Human Biochemistry*, 230–32.

[92] E. G. Krebs and E. H. Fischer, "Commentary on 'The Phosphorylase b to a Converting Enzyme of Rabbit Skeletal Muscle' by E. G. Krebs and E. H. Fischer," *Biochimica et Biophysica ACTA* 20 (1956): 150–157," *Biochimica et Biophysica ACTA* 1000 (1989): 297–309.

[93] See T. N. Raju, "The Nobel Chronicles, 1992: Edmond H. Fischer (B 1920) and Edwin G. Krebs (B 1918)," *Lancet* 355, no. 9219 (2000): 2004.

[94] A. De Lean, J. M. Stadel, and R. J. Lefkowitz, "A Ternary Complex Model Explains the Agonist-Specific Binding Properties of the Adenylate Cyclase-coupled b-Adrenergic Receptor," *Journal of Biological Chemistry* 255 (1980): 7108–17.

[95] See S. M. Foord, T. I. Bonner, R. R. Neubig, E. M. Rosser, J. P. Pin, and A. P. Davenport, "International Union of Pharmacology. Xlvi. G Protein-Coupled Receptor List," *Pharmacological Reviews* 57, no. 2 (2005): 279–88.

[96] See T. Hunter, "Discovering the First Tyrosine Kinase," *P.N.A.S.* 112, no. 26 (2015): 7877–82.

[97] J. A. Cooper and T. Hunter, "Changes in Protein Phosphorylation in Rous Sarcoma Virus-Transformed Chicken Embryo Cells," *Molecular and Cellular Biology* 347 (1981): 28.

[98] See, for reviews, S. R. Hubbard and J. H. Till, "Protein Tyrosine Kinase Structure and Function," *Annual Review of Biochemistry* 69 (2000): 373–98; M. A. Lemmon and J. Schlessinger, "Cell Signaling by Receptor Tyrosine Kinases," *Cell* 141, no. 7 (2010): 1117–34.

[99] See R. E. Dickerson and I. Geis, *The Structure and Action of Proteins* (New York: Harper & Row, 1969), 65.

[100] Collins, *Language of God*, 136–37.

[101] Haarsma and Haarsma, *Origins*, 51.

[102] See P. Johnson, *Darwin on Trial*, 2nd ed. (Downers Grove, IL: InterVarsity Press, 1993).

[103] G. G. Simpson, *The Book of Darwin* (New York: Washington Square, 1982), 17.

[104] T. Dobzhansky, "Nothing in Biology Makes Sense except in the Light of Evolution," *American Biology Teacher* 35, no. 3 (1973): 129.

[105] The following summary of Darwin's life up until the writing of *The Origin of the Species* is largely based on Simpson, *Book of Darwin*, 31–149.

[106] Mukherjee, S. *The Gene: An Intimate History*. New York, NY: Scribner, 2016.

[107] T. R. Malthus, *An Essay on the Principle of Population*, ed. Geoffrey Gilbert, 6th ed., Oxford World's Classics (Oxford: Oxford University Press, 2008; first published 1826).

[108] T. Davis, "Theistic Evolution: History and Beliefs," *BioLogos*, October 15, 2018, https://biologos.org/series/science-and-the-bible/articles/theistic-evolution-history-and-beliefs.

[109] Collins, *Language of God*, 200.

[110] W. Whewell, *The Bridgewater Treatises on the Power, Wisdom and Goodness of God as Manifested in the Creation. Treatise I-Viii.: Astronomy and General Physics Considered with Reverence to Natural Theology* (London: William Pickering, 1839), 356.

[111] Harry Emerson Fosdick, "Shall the Fundamentalists Win?," *Christian Work* 102 (June 10, 1922): 716–22.

[112] J. Polkinghorne, *Belief in God in an Age of Science* (New Haven: Yale University Press, 1998), 44.

[113] A. Gray, *Natural Science and Religion: Two Lectures* (New York: Charles Scribner's Sons, 1880), 108.

[114] L. M. Chiappe and L. M. Witme, eds., *Mesozoic Birds: Above the Heads of Dinosaurs* (Berkeley: University of California Press, 2002), 536 pages.

[115] C. Darwin, *The Origin of Species by Means of Natural Selection, or The Preservation of Favored Races in the Struggle for Life*, 4th ed. (London: John Murray, 1866), 367.

[116] See G. M. Erickson, O. W. M. Rauhut, Z. Zou, A. H. Turner, B. D. Inouye, D. Hu, and M. A. Norell, "Was Dinosaurian Physiology Inherited by Birds? Reconciling Slow Growth in Archaeopteryx," *PLOS One* 4, no. 10 (2009): e7390; N. Longrich, "Ancient Birds Flew on All Fours," Calgary: Wayback Machine, University of Calgary, 2017.

[117] W. Alvarez, "The Historical Record in the Scaglia Limestone at Gubbio: Magnetic Reversals and the Cretaceous-Tertiary Mass Extinction," *Sedimentology* 56 (2009): 137–48.

[118] P. Schulte, L. Alegret, I. Arenillas, J. A. Arz, et al., "The Chicxulub Asteroid Impact and Mass Extinction at the Cretaceous-Paleogene Boundary," *Science* 327, no. 5970 (2010): 1214–18.

[119] D. C. Johanson and Kate Wong, *Lucy's Legacy: The Quest for Human Origins* (New York: Harmony Books, 2009) 59–111.

[120] T. D. White, B. Asfaw, D. DeGusta, H. Gilbert, G. D. Richards, G. Suwa, and F. C. Howell, "Pleistocene Homo Sapiens from Middle Awash, Ethiopia," *Nature* 423, no. 6941 (2003): 742–47; T. D. White, B. Asfaw, Y. Beyene, Y. Haile-Selassie, C. O. Lovejoy, G. Suwa, and G. WoldeGabrie, "Ardipithecus Ramidus

and the Paleobiology of Early Hominids," *Science* 326, no. 5949 (2009): 75–86.

[121] See S. J. Gould and E. S. Vrba, "Exaptation—a Missing Term in the Science of Form," *Paleobiology* 8 (1982): 4–15.

[122] See S. Pääbo, *Neantherthal Man: In Search of Lost Genomes* (New York: Basic Books, 2014); T. M. Powledge, "Sexy Science: Neanderthals, Svante Pääbo and the Story of How Sex Shaped Modern Humans," *Genetic Literary Project*, March 6, 2014, https://geneticliteracyproject.org/2014/03/06/sexy-science-neanderthals-svante-paabo-and-the-story-of-how-lust-shaped-modern-humans/.

Made in the USA
Columbia, SC
10 October 2021

46646719R00124